T0138496

Model Behavior

Model Behavior

ANIMAL EXPERIMENTS, COMPLEXITY, AND THE GENETICS OF PSYCHIATRIC DISORDERS

Nicole C. Nelson

The University of Chicago Press CHICAGO & LONDON

The University of Chicago Press, Chicago 60637
The University of Chicago Press, Ltd., London

Published 2018
Printed in the United States of America

27 26 25 24 23 22 21 20 19 18 1 2 3 4 5

ISBN-13: 978-0-226-54592-9 (cloth)
ISBN-13: 978-0-226-54608-7 (paper)
ISBN-13: 978-0-226-54611-7 (e-book)
DOI: https://doi.org/10.7208/chicago/9780226546117.001.0001

Library of Congress Cataloging-in-Publication Data

Names: Nelson, Nicole C., author.
Title: Model behavior : animal experiments, complexity, and the genetics of psychiatric disorders / Nicole C. Nelson.
Description: Chicago ; London : The University of Chicago Press, 2018. | Includes bibliographical references and index.
Identifiers: LCCN 2017035137 | ISBN 9780226545929 (cloth : alk. paper) | ISBN 9780226546087 (pbk. : alk. paper) | ISBN 9780226546117 (e-book)
Subjects: LCSH: Behavior genetics—Research. | Animal models in research. | Mental illness—Animal models. | Mice as laboratory animals. | Behavior genetics—Methodology.
Classification: LCC QH457.N45 2018 | DDC 616.02/73—dc23
LC record available at https://lccn.loc.gov/2017035137

♾ This paper meets the requirements of ANSI/NISO Z39.48-1992 (Permanence of Paper).

CONTENTS

INTRODUCTION

A Furry, One-Ounce Human?

I am standing in the middle of a mouse colony room. On either side of me, shelves filled with clear plastic cages stretch from the floor almost all the way to the ceiling. Technicians in green hospital scrubs pull the cages in and out of the metal shelves like drawers as they move about on their morning rounds. Yellow note cards cover the front of each cage, labeled with cryptic notations indicating the age, sex, and strain of the mice, how many offspring they have had, and when the pups will be ready to be weaned. Behind one of the note cards, a female mouse and her silky pink newborn pups sleep in the corner of a cage, partially buried in the chips of corncob bedding that cover the plastic cage floor. This mouse has already finished her run as an experimental subject, but in a few weeks her pups will join nearly a hundred other young mice from this colony room in the next round of a study on the genetics of alcohol addiction.

This particular mouse room is not especially remarkable. In many ways, it is just like dozens of other mouse rooms that I visited on my travels through animal behavior genetics laboratories. Yet standing here, for a moment I let myself drift out of seeing this space as part of the ordinary backdrop of contemporary biomedicine, and I am struck by the strangeness of this enterprise: thousands of mouse rooms like this one, containing millions of mice, all across the United States, all across the world, all built in hopes of better understanding the human. How did we arrive at this place, where so many of our hopes for understanding human biology are concentrated on this small organism? And how do we expect to extract knowledge about our own addictions and

FIGURE 1. Mouse housing room with ventilated cages at a university in Canada—the first mouse laboratory I visited as part of this project. Photograph by the author, July 2006.

anxieties from those tiny pink mice in the cage that are not yet able to even open their eyes?

Animal experiments are a crucial part of the machinery of contemporary biomedicine, and rodents are the most widely used organisms in biomedical research today—so much so that Angela Creager, Elizabeth Lunbeck, and Norton Wise (2007, 1) argue that "at the dawn of the twenty-first century, the face of biology may well be that of a laboratory mouse." A survey of publication trends in the journal *Genetics* shows a concentration in genetic research on a limited number of species over the past fifty years. In the 1960s, the number of studies published in the journal on research with so-called model organisms was about the same as publications on other organisms. But by the close of the twentieth-century model, research on model organisms accounted for more than 75 percent of the journal's publications (Dietrich, Ankeny, and Chen 2014). A study of all scientific publications indexed in the National Library of Medicine's database shows a similar trend. The number of papers published annually using mice and rats has more than quadrupled since the 1960s, and by 2009 the mouse alone was the subject of three times more publications annually than yeast, fruit flies, zebrafish, and nematodes combined (Engber 2011). Of all the research rodents, the most popular strain is a mouse with a dark brown coat called the C57 Black 6 (C57BL/6), first bred by C. C. Little in the 1920s at the Jackson Laboratory (Rader 2004; see also Gaudillière 2001). Although precise figures on the specific mouse strains used in research are difficult to obtain, the C57BL/6 easily outsells all others, accounting for between half and two-thirds of mouse sales by some estimates (Engber 2011). Taken together, these figures highlight the considerable amount of biomedical knowledge that is generated based on experiments with rodents—or even with one single strain of mouse.

Mice appear everywhere in biomedical stories about ourselves as humans. If you examine a newspaper article about a new medical finding, chances are good that you will find a mouse lurking in the evidentiary basis for the breakthrough. A *New York Times* article on the relationship between exercise and anxiety, for example, starts out describing in general terms how exercise impacts brain development, before revealing the source of the new information: a series of experiments with laboratory mice (Reynolds 2013). Researchers at Princeton gave running wheels to one group of mice and then compared them to another group of mice that had no exercise equipment. They found that the runners were both less anxious and had more new neurons in their brains. What all of this suggests, the article reports, is that activity reshapes the brains of animals in ways that dampen the effect of stressful situations. "Of course, as we all know, mice are not men or women," the article cautions in the penultimate paragraph. But the last word is given to the lead researcher on the aforementioned mouse studies. He concludes: "It's not a huge stretch to suggest that the [brains] of active people might be less susceptible to certain undesirable aspects of stress than those of sedentary people."

As much scholarship in animal studies has shown, humans have long used animals to understand themselves. John Berger's (1980) influential and wide-ranging essay "Why Look at Animals?" laid the groundwork for a scholarly agenda of examining animals as symbols, as companions, and as spectacles. The "parallelism of their similar/dissimilar lives," as Berger (1980, 7) put it, has throughout history provided fertile ground for humans to craft narratives about themselves. Science and technology studies scholars have similarly examined how scientists use animals to produce narratives about humans. Donna Haraway's (1989) seminal work *Primate Visions* examined how primatology was mobilized in a wide variety of scientific and popular settings to generate stories about the naturalness of particular human social arrangements. From dioramas of primate "families" in natural history museums to laboratory research on the mother-infant bond, Haraway showed how nonhuman primate science reflected and reinforced Western notions of gender and race. Rebecca Lemov's (2005) analysis of behavioral experiments with rodents similarly showed how these experiments captured the popular imagination in North America in the 1960s because they aligned with Cold War visions of improving the human through behavioral engineering.

Using mice as stand-ins for humans in scientific experiments has today become so commonplace that news articles such as the one mentioned above can effortlessly slide back and forth between the animal and the human with only a hint of a caveat. It has become easy to see the mouse, as one veterinarian put it to me, as simply a "furry, one ounce human"—but one that exists within the exquisitely controlled setting of the laboratory. Laboratory mice straddle the boundary between the natural and the artificial, a property which numerous commentators on mouse research have suggested is partly responsible for their success as research tools (Davies 2013; Rader 2004; Sismondo 1997). The mouse's status as a living being that shares an evolutionary history with humans imbues it with the epistemic authority of the natural world, and the way that it has been altered to function as a scientific tool gives researchers opportunities to design experiments that would be impossible with human subjects. The abundant supply of mouse strains available to researchers today provides animal behavior geneticists with an army of genetically identical individuals. These identical mice can be tested again and again in different conditions to see, for example, how many times they will seek out a bottle of alcohol and take a drink, or venture out into the unprotected spaces of a maze. Mice can be selectively bred to see if mating one high drinker with another will produce offspring with an even greater affinity for alcohol. Or they can be manipulated at the molecular level to see if removing a particular gene from their genome changes their behavior. Their brains can be probed, gene expression levels tracked, and the relationship between genetic markers and particular traits statistically computed—all to gain insight into the inherited risks and chains of molecular events that might work together to produce behavioral pathologies in humans.

Historians of science have repeatedly demonstrated, however, that the rise to prominence of model organisms such as the mouse was far from inevitable, no matter how compelling their advantages as research tools may now seem (Ankeny 2001; Creager 2001; Gaudillière 2001; Kohler 1994; Logan 2002; Mitman and Fausto-Sterling 1992; Rader 2004). Jean-Paul Gaudillière (2001) and Karen Rader (2004) have argued that the very ubiquity of the mouse in contemporary biology obscures the institutions, social practices, and conceptual infrastructure that support its widespread use. If you ask an animal researcher why the mouse is a good model for studying human disorders, she will likely reply with some familiar arguments: they are inexpensive and easy to maintain in the laboratory; they reproduce quickly and have a short life-span; their anatomy, genetics, and behaviors are well characterized; and they share many biological and genetic similarities with humans. But this neatly packaged rationale decontextualizes the mouse from the places and circumstances where it came to be seen as a good tool for investigating biomedical questions. Gaudillière and Rader describe how institutional and political support for mouse breeding programs and commitments to genetic theories of causation in cancer research facilitated the mouse's transformation into the standard laboratory organism that it is today. As Rader (2004, 15) puts it, it is more productive to think about the laboratory mouse and other model organisms as "the result, rather than the cause, of consensus amongst early twentieth century experimental biologists."

This book travels inside the animal behavior genetics laboratory to examine how animal behavior geneticists continue to build and maintain the infrastructures supporting mouse models as knowledge production tools. While the institutions that supply researchers with inbred mouse strains and the rationale for using mice as models for human disorders have been in place for decades, the presence of these established infrastructures does not mean that researchers no longer have to argue that studying a mouse in a maze is a useful way of producing knowledge about human anxiety. In their discussion of mouse models of Alzheimer's disease, Lara Huber and Lara Keuck (2013) describe the establishment of biomedical animal models as a continuous and contested process. They write that "the representational relation between model and target is subject to ongoing validation, furthermore that validation in biomedicine is volatile and that established animal models require continuous reassessment" (386). Jamie Lewis and colleagues (2013, 780) similarly argue that the capacity of mice to model human disorders is something that "need[s] to be achieved . . . their representational capacities have to be worked at continually." Mette Svendsen and Lene Koch (2013) have also examined how the bodies of laboratory animals are continually made and unmade into the kinds of bodies that can substitute for obese humans or preterm infants.

What are the means by which animal behavior geneticists "achieve" the ongoing validity of their models? What exactly do these researchers believe their models are (or are not) useful for, and how do they manage the strength of the associations they make between animal and human, behavior and gene? Do they use different techniques to accomplish validity in the eyes of scientists outside of the field, funding agencies, or the general public? In exploring how researchers weave evidence together to create a rationale for engaging in animal work, we will see that researchers draw on more than just biological similarities between mice and humans. They also use cultural knowledge about psychiatric disorders, and knowledge about environmental effects gleaned through the hands-on experience of working with animals. Mice and humans, mazes and drugs, genes and behaviors, practical experience and widely recognized scientific findings—all these are continually and carefully set in relation to each other to create a space that functions as a credible site for producing knowledge about human behavior.

Animal behavior genetics is a good place to observe these kinds of processes because the field is positioned within multiple debates of widespread social salience: those concerning the nature of human behavior, the meaning of genes, and the value of animal research for addressing pressing biomedical problems. These debates make the accomplishment of model validity especially challenging. Although the mouse may be widely used as a model for understanding human disorders, this approach is certainly not without its detractors. The *New York Times* article described above may glide easily between mouse and human, but the comments section hints at the objections to this approach lurking under the smooth surface of the article. One skeptical reader, taking issue with the article's headline—"How Exercise Can Calm Anxiety"—retorted sarcastically, "I now append 'in mice' to all the headlines I read about amazing medical news." Animal rights groups routinely question the benefit that animal research provides in understanding human health. Researchers in the pharmaceutical industry have also begun to question the utility of animal research, pointing out that even well-validated and widely used models have not necessarily led to successes in clinical trials or advances in drug development (Dawson and Tricklebank 1995; Geerts 2009). Even high-ranking scientific leaders such as Francis Collins (2011), the director of the National Institutes of Health, have suggested that it may be time to consider skipping animal model assessments of drug efficacy altogether in programmatic calls for reimagining the translational research enterprise.

The field is also unique because its central premise—that there are heritable components to human behavior—has repeatedly generated heated public controversy. The history of behavior genetics has been punctuated by several highly public debates about the racist undertones and discriminatory policy implications of research on heredity and intelligence. Studies on human genetic markers linked with homosexuality have similarly sparked public interest in and criticism of the field's research. Aaron Panofsky (2014) has shown how practitioners' attempts to deal with these persistent controversies and maintain their legitimacy has fundamentally shaped how behavior genetics operates as a knowledge producing field, creating an unusual situation where the field's authority and methods are repeatedly called into question.

Finally, establishing plausible relationships between mouse experiments and human disorders is arguably especially challenging in animal behavior genetics, where researchers use animals to gain insight into disorders that even they themselves sometimes describe as "uniquely human." Although mice offer many advantages as research tools, they cannot replicate many core features of behavioral disorders. Mice cannot lose their jobs or damage their relationships because of their excessive drinking, and they cannot talk about their subjective experiences of the paralyzing dread of anxiety and how drugs alter it. Human behavioral disorders such as alcohol addiction and anxiety are themselves heterogeneous and ill-defined, making it difficult to say how well mouse models represent these elusive targets. And because mice have been so dramatically altered to serve particular functions in laboratories, it is unclear how much of the knowledge produced with mice is an artifact of these instrumental manipulations. This paradoxical nearness and distance between the mouse and the human, the need to pull them together while maintaining an awareness of their fundamental differences, animates many of the dynamics of modeling work that I explore in what follows.

ANIMAL EXPERIMENTS AND COMPLEXITY

The distinctive epistemic culture that has emerged under these conditions offers opportunities for asking additional questions about what laboratory work looks like when researchers believe that both the human behaviors they are studying and the mice they are using to model them are "complex." It is easy to look at an experiment using a lone mouse in a cage as a means to study alcoholism and see it as irredeemably reductionist—at least, that is how I saw such experiments when I first started this project. In my initial interviews with animal behavior geneticists, I peppered them with questions about the validity of their models and their relationship to human psychiatric conditions: How could studying individual animals in isolation capture the social dynamics that are so important to drinking? Did the fact that an animal model resembled a human behavior really ensure that it would be useful for finding new pharmacological treatments? In retrospect, I am not sure what I expected to happen when I asked these questions—did I think that they would debate me, or throw up their hands and exclaim, "You've found us out!"?—but I do know that I was not expecting the response I received. Far from finding these lines of inquiry irritating or threatening, they encouraged me along, telling me that I was "on the right track" and "asking exactly the right questions." For every problem that I could spot with their models, they offered two more. Moreover, they seemed to relish conversations about all the ways in which their models were limited. I felt outwitted at my own game.

Gradually, I began to move away from seeing my project as one of pointing out complicating factors that animal modelers had ignored, and instead, I became interested in how researchers dealt with all of the complicating factors they themselves kept in view. It seemed to me that they had created an impossible situation for themselves. Rather than taking an expansive, "big data" approach to studying behavior and aiming to smooth out uncertainties or problems of measurement through sheer volume, they opted to study behavior in the controlled setting of the laboratory. But they also believed that the creation of a truly controlled laboratory environment, where all of the parameters that might impact behavior were known and accounted for, was presently out of reach. Under such conditions, researchers saw their capacity to produce stable associations between genes and behavior as substantially constrained. A conversation I had with one animal behavior geneticist solidified this point. He told me he thought I was being "too generous" in my evaluation of the particular model we were discussing: "I tend to think of any behavioral model as more of a pudding skin over top of a molten morass," he said. "You know it's going to give at some point, you just don't know when." How could research move forward where there was such deep ambivalence in the scientific community, even within its core set of practitioners and even about its established methods? This book is an attempt to understand what laboratory work looks like under these conditions.

In focusing on knowledge production under expectations of complexity, I aim to contribute to two additional literatures: social studies of genomics, and laboratory studies. Much has been written about the limitations of genetic approaches to understanding human problems, and also the potential harms that might result from directing too much funding and public attention to these approaches. The heavy investments in genetics made by commercial entities and public institutions in the closing decades of the twentieth century inspired much critical commentary on the value of genetic research. Analysts argued that these approaches had the tendency to flatten complicated causes of disease, focusing attention on genes as the primary causal factors for human differences and disorders while pushing social contributors to health and illness aside (e.g., Alper and Beckwith 1993; Conrad 1999a, 1999b; Conrad and Weinberg 1996; Nelkin and Lindee 1995). Abby Lippman (1992) coined the term *geneticization* to describe the growing tendency for scientists, medical professionals, and laypeople to attribute human differences to genetics, with potentially dangerous implications for social policy. Commentators both inside and outside of the sciences were especially critical of genetic approaches to studying behavior. Scientists Richard Lewontin and Steven Rose have both been outspoken critics of what they see as the reductionist fallacies inherent in looking for "genes for" behavior (Lewontin 1991; Lewontin, Rose, and Kamin 1984; Rose 1997). Sociologist Troy Duster (1990, 2003) has expressed similar concern about methodological issues in and the social dangers of behavior genetics research. Duster (2003, 163) warned that efforts to investigate poorly defined behavioral categories using molecular methods "could easily become the phrenology of the twenty-first century."

Following hot on the heels of these analyses of geneticization and reductionism, however, was a cluster of books and articles questioning the prevalence of these frameworks. In his study of genetic research on schizophrenia, Adam Hedgecoe (2001) argued that the dominant narrative psychiatric geneticists presented in their review articles was not one of geneticization, but what he called enlightened geneticization—a discourse that acknowledged the importance of both environmental factors and genetics. Nikolas Rose (2007) argued that critics largely misunderstood the nature of twenty-first-century biology. The "new molecular knowledges of the human condition" (51), he argued, were not deterministic but probabilistic, aiming to characterize susceptibilities and make them open to intervention rather than pronouncing on fates. In his words, these new biological explanations were, "to use a much abused term, 'complex'" (51). Hallam Stevens and Sarah Richardson (2015) have described invocations of complexity as a hallmark of postgenomic research. "As scientists narrate the history of genome sequencing projects," Stevens and Richardson write, "they trace a path from a simplistic, deterministic, and atomistic understanding of the relationship between genes and human characters toward . . . an emphasis on complexity, indeterminacy, and gene–environment interactions" (3).

Determining whether postgenomic biology in general or animal behavior geneticists in particular have truly embraced complexity is, I will argue, not really an answerable question; but we can examine how scientists cultivate complexity talk, and how they use it to promote specific styles of experimenting and claiming. Analysts have taken up something like this research agenda in emerging fields such as epigenetics (see, e.g., Niewöhner 2011), but we presently lack a critical mass of studies examining the relationship between assumptions of complexity and experimental practice in more established areas of the life sciences. In the laboratories I studied, for example, researchers used discussions about complexity not to completely reimagine their research programs but to direct attention to the laboratory environment and its impact on animal behavior. This seemingly subtle shift, as we will see, had important effects on the knowledge researchers gained from their work with animals. On the one hand, seeing how particular strains of mice would consistently drink from the alcohol bottles that other strains refused to touch provided convincing experiential evidence that there was some inherited propensity to drink. But on the other hand, seeing how noise from construction projects or an unexpected fire alarm in the research building could change a mouse's drinking patterns showed researchers that the effects of these inherited factors were far from straightforward or deterministic. As a result, even though the animal behavior geneticists I studied remained professionally committed to the search for genetic factors influencing behavior, their commitments to complexity meant that in practice, they spent much of their time identifying and managing environmental factors influencing behavior.

Examining the relationship between assumptions of complexity and experimental practice is also important for laboratory studies. One of the early projects for laboratory studies was to investigate how local processes of scientific work gave rise to supposedly universal truths—or, as Susan Leigh Star (1985, 391) put it, the techniques by which the "local uncertainties encountered by working scientists [are transformed] into global certainty." Scholars were intrigued in particular by how scientists themselves, who were keenly aware of the labor involved in getting experiments to work, nonetheless came to regard their findings as laws of nature. Harry Collins (1985) described the transformation between these two stances as a "crystallization of certainty" that took place on the completion of an experiment. He wrote: "Scientists and others tend to believe in the responsiveness of nature to manipulations directed by sets of algorithm-like instructions. This gives the impression that carrying out experiments is, literally, a formality. This belief, though it may be suspended at times of difficulty, re-crystallizes catastrophically upon the successful completion of an experiment. . . . Doubts, if they arise, last for only a short time" (76).

Like elastics under tension, the scientists in Collins's description snapped back at the first opportunity to the more comfortable position of attributing agency in the experimental setting to nature, and seeing the results of the experiment as fact. Bruno Latour and Steve Woolgar (1979) portrayed this process as more gradual. In their seminal study *Laboratory Life*, they traced the arc of one particular fact established in the laboratory of Nobel Prize winner

Roger Guillemin: that "TRF (thyrotropin-releasing factor) *is* Pyro-Glu-His-Pro-NH2" (147). They described the credibility of such claims as increasing or decreasing on an ongoing basis "rather like the daily Dow Jones Industrial Average" (17), but they too identified an inflection point where statements gained enough credibility that it became too costly to question their reality. What was once merely a conjectural statement, they argued, became a "split entity" (176) that appeared to contain both an independent object and a statement about that object.[1]

The kinds of scientific communities and processes that early laboratory ethnographies investigated were carefully chosen for their analytical leverage: they provided hard cases for demonstrating that "facts are thoroughly understandable in terms of their social construction," as Latour and Woolgar (1979, 107) put it. If analysts could show that even the activities of physicists or straightforward statements about the structure of a protein were amenable to constructionist analysis, then that analysis would surely hold in other scientific fields where social processes were more evident. But not all scientists expect that nature will respond with "algorithm-like instructions" to their experimental manipulations, or aim to establish the kinds of facts that Guillemin's laboratory sought. The laboratories that I studied seemed quite unlike the authoritative "truth spots" (Gieryn 2002, 2006) they are typically portrayed as: even these exquisitely controlled spaces were full of epistemic problems to be managed, problems that allowed even seemingly stabilized results to disappear. In my observations, moments of "catastrophic recrystallization" never seemed to arrive for me to unpack them. Researchers treated behavior as something more capricious that could change depending on the details of the experimental protocol, the time of day, or the person conducting the experiment. It felt as though researchers were living in an extended moment of doubt and uncertainty, much like the moment that Collins (1985) described as preceding the successful completion of an experiment.

As Hugh Gusterson (2008, 559) might put it, the knowledge produced through animal modeling work seemed to be "not just constructed, but hyperconstructed." In Gusterson's study of contemporary nuclear weapons science, he argued that treaties banning nuclear weapons testing have created a lack at the epistemic core of the field, one that has made it difficult for debates to settle and for "normal science" to take hold. In the absence of the specific kind of evidence that weapons scientists desire—data from actual field tests of weapons—even experts in the field continue to disagree about the reliability of new weapons designs and the simulations intended to supplant physical testing. A similar kind of epistemic void plagued the animal behavior genetics labs I studied. The very description of animal experiments as "models" for human disorders marked them out as substitutes, proxies for phenomena of interest that were inaccessible in particular ways to researchers. The alignment between the animal model and the human disorder was one that researchers approached with a sense of pragmatism, arguing that their animal models were "good enough" to serve particular purposes (Lewis et al. 2013), even if they were not ideally suited to the tasks for which they were enrolled. The human hands conducting the mouse experiments never seemed to be fully erased from the finished result, and practitioners operated under the expectation that the knowledge they produced was partial, inadequate, and subject to ongoing revision.

Studies of laboratories like these are needed because without such studies we might mistake the specific kinds of facts and process that Latour, Woolgar, Collins, and others have described for universal features of laboratory work. In addition to studies examining how scientists produce facts that are "enduring, translocal, [and] transtemporal," as Park Doing (2008, 286) has put it, we also need studies examining how laboratory scientists create knowledge that is partial and situated. Without such studies, it is much less obvious why scientists would want to reduce the scope of the claims they make, or point out that their results hold only for particular situations—both moves that we will see in examining how animal behavior geneticists carry out their work. Examining knowledge production under assumptions of complexity also makes more visible what I call the "extrafactual" work of the laboratory—work done to validate models or understand laboratory environments that is not specifically aimed at producing experimental data, but supports and shapes this process. By taking these other knowledge production processes and products into account, I argue that we gain a more comprehensive understanding of how mice function as models, and of what experimental work is capable of producing more generally.

To articulate these ideas, I draw on long-standing constructionist analogies in science and technology studies (STS). For example, I describe how researchers negotiate the capacities of animal models as a process of building "epistemic scaffolds" to support future fact construction, and how researchers gain "epistemic by-products" in the process of producing genetic knowledge about psychiatric disorders. This constructionist language is somewhat at odds with tendencies in animal studies to emphasize the liveliness of animals, and recent trends in STS toward new materialist thinking (see, e.g., Barad 2003, 2007). Constructionist and new materialist approaches tend to place different emphasis on human agency versus the agency of the material world:

while one emphasizes the role of scientists' mouse experiments in constructing human cultures, the other focuses on the mouse's own natural history, actions, and needs in the laboratory setting.

These are tensions that I am aware of but do not attempt to resolve in this book. As philosophies, constructionism and new materialism may not be as incompatible as they first appear. Diana Coole and Samantha Frost (2010, 27) write in their essay on the new materialisms that "it is entirely possible . . . to accept social constructionist arguments while also insisting that the material realm is irreducible to culture or discourse and that cultural artifacts are not arbitrary vis-à-vis nature." But as analytical approaches, they are responding to very different matters of concern. Language emphasizing the liveliness of the natural world is a useful way of counterbalancing scholarship that has treated animals as blank slates on which humans create their narratives, or that has argued more generally for treating the world as text. Vinciane Despret's (2004, 131) description of animal research as a process of "mutual articulation" is a compelling example of work that attempts to transform a world full of "enthusiastic automata observing strange and mute creatures" into a world full of interactions between scientists and animals. But the human-centered affordances of constructionist metaphors have done important work for STS in counteracting realist epistemologies that claim that the scientific method simply lets nature speak. As laboratory ethnography has shown, it is altogether too easy for human agency to be erased from the scene of the experiment, especially after it has concluded.

I find constructionist language compelling for thinking about animal modeling because of what Sergio Sismondo (2008, 14) identifies as some of the core features of constructionism: its insistence on treating science as an active process, and the products of science as "not themselves natural." These are useful correctives in a situation where both scientists and analysts tend to talk about animal models in ways that disguise aspects of the active modeling process. The ways that many analysts have talked about animal modeling, for example, creates the impression that findings are first worked out in model systems and then moved to new contexts, thereby obscuring the interactions between the model and the modeled disorder that take place at all stages of this work. Carrie Friese and Adele Clarke's (2012) description of how experimental findings and animals are "transposed" into new settings or Angela Creager's (2001, 5) description of model organisms as "exemplars for studying and understanding other entities and organisms" are both extremely useful vocabularies for thinking about processes of circulation, but less so for thinking about processes of developing models. Using a constructionist framework to analyze the process of model development is one way of drawing attention to that activity. Likewise, constructionist language helps guard against the temptation for both actors and analysts to ask realist questions about models such as, is the behavior that we are seeing in this model *really* alcoholism? The assertion that animal models of alcoholism and indeed alcoholism itself are not themselves natural serves as a reminder that there is no human-independent position from which to make such evaluations. While constructionism may not succeed in highlighting all the various agencies at work in the experimental scene, it still offers a valuable analytical vantage point from which to view animal modeling work.

ENTERING THE ANIMAL BEHAVIOR GENETICS LABORATORY

To better understand how researchers use mouse models to generate knowledge about the genetics and neurobiology of human behavioral disorders, I spent time at a university that I will refer to here as Coast University. The university was situated in a hilly city on the West Coast of the United States, and the laboratories in the Department of Neuroscience were spread out in many buildings across campus and in the buildings of the affiliated teaching hospital. My host during my time at Coast was Dr. Daniel Smith, a senior researcher in the neuroscience department who was well known for his work in addiction genetics. Dr. Smith had made his way into addiction research via training in psychology and behavior genetics, and with the exception of a brief stint at a pharmaceutical company after graduate school, he had spent the entirety of his career in academic research. He had also spent nearly the entirety of his research career working with rodent models. Although like many in the field Dr. Smith had developed an allergy to mice that made it difficult for him to keep doing animal experiments, he talked with great familiarity and affection about the "little bitty guys" he had spent so many decades working with.

The Department of Neuroscience at Coast was known for its strengths in drug and alcohol research, and broadly speaking, the aim of the research there was to identify genetic and neurological mechanisms of susceptibility to addiction or sensitivity to particular drugs. Researchers were also interested in related "behavioral traits" such as anxiety and impulsivity, because these traits often coexisted in people with alcoholism and were thought to contribute to the risk of addiction or relapse. A person experiencing anxiety, for example, might use alcohol to relieve their symptoms, putting them at greater risk for an alcohol use disorder. Coast researchers used a variety of techniques to carry out their research agenda, from what they described as "old school" methods that had been in use for decades (such as selective breeding, where researchers mate mice with high or low scores on a particular test to see if those behavioral tendencies are inherited by the offspring), to cutting-edge techniques requiring expensive and high-tech equipment (such as microdialysis, where researchers use a small tube inserted into the mouse's brain to measure neurotransmitter levels in a specific brain region as the mouse is doing a behavioral test).

The work taking place at Coast was somewhat difficult to categorize using a basic/applied science dichotomy. Although the researchers there described their work as "basic science," it had a strongly clinical orientation. Their ultimate aim was to generate information that would be useful in studying human behavioral disorders, but they held onto the basic science label because they believed that the application of their work was a long way off, and because they thought it was unclear how exactly their findings might translate into clinical uses. Their research was almost entirely funded by public institutions such as the National Institute on Alcoholism and Alcohol Abuse (NIAAA), the National Institute on Drug Abuse (NIDA), and the Department of Veterans Affairs Office of Research and Development. None of the researchers at Coast had active relationships with the pharmaceutical industry that I was aware of, although several graduates of the program had taken research positions within pharma companies. In general, the researchers I observed seemed quite financially secure. Like all scientists, they were continually engaged in writing and resubmitting grant proposals, but financial constraints did not figure prominently into the decisions made at laboratory meetings, and many of the laboratories in the department were well funded enough that they employed several full-time staff. Dr. Smith, for example, employed two lab managers and four full-time technicians in addition to his postdocs and graduate students.

The Smith Laboratory specialized in developing and troubleshooting mouse models of addiction, while other laboratories in the department specialized in identifying specific genes or brain regions involved in addiction, or in studying the mechanisms of action of those genes or neural pathways. Several research groups at Coast worked quite closely with the Smith Laboratory, and the activities taking place in their laboratories shared a good degree of overlap. The relationships between these researchers were long-standing and affectionate, and as the most senior researcher, Dr. Smith served as something of a patriarch in this close-knit group. He had mentored many of the researchers in the department at various stages in their careers, and over time, these mentorships had turned into long-term collaborations. For example, Dr. Smith shared a number of ongoing projects with Dr. Laura Martin, another senior researcher in the department. The two shared a ritual, built over a number of years, where they lunched together once a week and talked science. Dr. Smith seemed happy to welcome into the fold all those who expressed interest in the type of work that he did, and I surely benefitted from his openness and mentorship support: not only did he welcome me into his laboratory and encourage others to take me seriously as a scientist of a different stripe, he even went so far as to help me secure a conference registration fee grant so that I could see more of the broader behavior genetics world.

As is likely evident already, my research follows in the tradition of laboratory ethnographies, where analysts spend time as participant observers in laboratories to gain insight into how the day-to-day work of knowledge production unfolds (Knorr 1981; Knorr-Cetina 1999; Latour 1987; Latour and Woolgar 1979; Lynch 1985; Traweek 1992). I became interested in mouse research after reading a magazine article that asked whether these animals could be trusted as models for human behavior (Yeoman 2003). In summer 2006, I started interviewing researchers who had published on methodological issues in behavioral animal modeling. In a telephone interview, Dr. Smith answered all of my questions with what I would come to know as his characteristic generosity and good humor. Whereas other researchers I interviewed wondered why a "social scientist" was not interested in bioethics or was inquiring about the details of their methods, Dr. Smith seemed entirely unfazed by my interest in the nitty gritty of laboratory work. He readily agreed to let me come tour his laboratory some months later, where he treated me like the other visiting scientist that he hosted in his laboratory, meeting with me about my research interests and instructing his staff to show me their facilities and ongoing experiments. In these first few days, I scrambled to keep up with the technicians while they changed in and out of their scrubs at lightning speed, and held my breath while I watched for the first time as a researcher cut open a euthanized mouse. After returning from this trip, I sent Dr. Smith a proposal for doing a longer term of fieldwork under his supervision and returned to Coast in January 2008.

While at Coast, I involved myself as much as possible in the rhythm of daily life in the laboratory. I focused in particular on moments where discussions of methodological issues might be especially prominent, such as courses, training sessions, or setting up new experiments. I took the introductory behavior genetics class, co-taught by Dr. Smith and his colleague Dr. Ruth Tremblay, with the new group of graduate students in the behavioral neuroscience program. I also took the mouse handling course for new laboratory practitioners, where Aiden from the animal care staff patiently walked me and my shaky hands through the process of injecting my first mouse with saline solution. I attended the weekly laboratory meetings where recent experimental results were discussed and new meetings were planned. Dr. Smith graciously incorporated me into his weekly meeting schedule for his graduate students and postdocs, where we discussed events taking place in the laboratory as well as general topics such as the manuscript review process, media reporting on scientific research, and past and future projects for his laboratory. While in the lab, I shadowed graduate students, postdocs, and technicians as they conducted their experiments; and I often acted as the "scribe" for these experiments, writing down figures such as mouse weights and quantity of alcohol consumed while the researchers went about other tasks. I filled bottles with alcohol solution, labeled test tubes for genotyping tissues, prepared mouse cages for upcoming experiments, scored behavioral data, and proofread data sheets. I spent a good deal of time chatting at the lunchroom table, the heart of the Smith Laboratory, where grad students and technicians tended to congregate before and after experiments. I also interviewed a number of researchers, graduate students, and staff. The methodological appendix provides more detail on my fieldwork, interviewing, and data analysis methods (including a complete list of individuals that I interviewed and their institutional and disciplinary locations), but one more methodological point deserves a note here: The majority of the names I use, including the names of field sites, are pseudonyms. I use real names only when I refer to the published literature or quote from an interview with someone who agreed to be identified by name and was not associated with an anonymized field site. I discuss in more detail in the appendix the reasons for conducting this fieldwork under the promise of anonymity.

In addition to this fieldwork at Coast University, I visited numerous other laboratories in the United States, Canada, and Europe. These visits were similar in format to my initial visit to the Smith Laboratory, often including a laboratory tour and some observations of ongoing experiments, and interviews with researchers, students, and sometimes technicians at that site. The majority of these visits and interviews were conducted between 2008 and 2009. I also gathered information on a group I call the Alcohol Research Group (ARG), one of several consortium projects funded by the NIAAA. It had been running for approximately eight years at the time I started following it, and brought together members from about twenty laboratories from across the United States (including members from Coast University) to work on a set of common animal models. I observed one of the annual meetings of the ARG in June 2008 and interviewed about a third of the researchers involved in the project, focusing on researchers who were working primarily with mouse models. Finally, I attended several conferences and workshops as a participant observer, such as annual meetings of the International Behavioural and Neural Genetics Society (IBANGS) and the Behavior Genetics Association (BGA), as well as Measuring Behavior, an interdisciplinary conference focused on methods, techniques, and tools for studying behavior. More details on these additional field sites and the interviews conducted there are again available in the appendix.

Although these additional laboratory visits and interviews sometimes offered informative contrasts to my fieldwork at Coast, they were not intended to be comparative or to contribute to a more complete picture of the animal behavior genetics field. The researchers I interviewed and sites I visited outside of Coast were, in many respects, extensions of Coast's epistemic community—in searching out people who were also interested in animal behavior genetics methodology, I often found people who had similar disciplinary backgrounds and opinions to those that I interacted with at Coast. This became abundantly clear during one interview with a Canadian researcher: I told him that I had heard quite a lot about a particular methodological issue, and he responded that was only because I was in the "echo chamber" with other people who cared about methods. The thoughts and opinions of the researchers I interacted with at Coast and elsewhere should thus not be read as standing in for commonly held positions in animal behavior genetics. I do not want to give the misleading impression that methodological preoccupations are a defining characteristic of the field as a whole. My aim is to use this particular group of scientists and their unusual attention to methods as a site for investigating the questions laid out previously, such as how researchers who believe that psychiatric disorders are complex negotiate the relationship between animal and human, and the solidity of their findings. In a sense, I am using this group of researchers as my own "model system" for exploring how knowledge production proceeds under a specific set of aims and assumptions.

PLAN OF THE BOOK

This book is divided into three parts. The first part examines what researchers at Coast meant when they described psychiatric disorders and animal models as "complex," and how this talk contributed to the formation and maintenance of their epistemic community. In the first chapter, I introduce a series of difficulties animating their experimental practice, each of which stemmed from what researchers described as the complexity of behavior: the problem of using one complex entity to model another, the difficulties of identifying and controlling for the many environmental inputs into behavior, and the appropriate strategies for breaking down complex phenomena into small-enough units for study. The combined force of these layers of epistemic difficulty resulted in what I call a "complexity crisis" for beginning practitioners, where they began to doubt that anything could firmly be said about genes and behavior at all. When the dust settled on this moment of crisis, practitioners emerged with revised expectations about the kind of knowledge they would be able to produce.

Chapter 2 continues the discussion of how researchers used narratives about complexity to shape expectations about knowledge production. Researchers at Coast used narratives about past methodological controversies, such as failures to replicate the result of knockout experiments in the 1990s, to craft an image of how a careful practitioner should craft scientific claims. Connecting these histories to moments of controversy around heredity, race, and intelligence lent further weight to Coast researchers' admonitions to formulate scientific claims cautiously. They used these past narratives to project an imagined trajectory for the field's future, one where the pathway to stable associations between genes and behavior was long and uncertain.

With this characterization of Coast researchers' stances on knowledge production in place, part two examines how they proceeded with experimental work. Under assumptions of complexity, other kinds of extrafactual knowledge production become more visible. Chapter 3 examines researchers at work on the epistemic foundations of their field, building up and negotiating claims about what kind of knowledge animal models can produce. Using the example of the "elevated plus maze," a rodent model for anxiety, I develop the metaphor of an "epistemic scaffold" to illuminate how scientists created and contested claims about the utility of animal models. The function of a scaffold as a platform for doing work highlights the importance of the methodological work that takes place prior to the production of specific facts, and the flexible, temporary nature of scaffolding draws attention to the active processes of building up and breaking down these epistemic claims.

Chapter 4 examines another knowledge production process taking place alongside the production of facts in the laboratory—the accumulation of knowledge through the process of conducting laboratory work. The "epistemic by-products" that Coast researchers generated as they attempted to produce genetic findings constituted an important part of their understanding of behavior. Knowledge about how the laboratory environment impacted mouse behavior acted as a counterbalance to genetic knowledge, and researchers used this knowledge deliberately to shape others' experimental practices, adjust the value of their claims, or ground genetic findings in their sites of production.

The final section of the book uses the arguments developed throughout to examine how animal modeling contributes to scientific and cultural understandings of psychiatric disorders. Chapter 5 uses an exploration of extrafactual work of animal modeling to reveal the interpretive flexibility of the models. While the model of binge drinking I discuss may look from some vantage points like it belongs squarely in a long line of reductionist research and medicalization of public problems, I show how researchers' work with the model made other understandings possible. The success of using manipulations to the mouse's environment to engender heavy drinking, for example, made it possible for researchers to think about and talk about human drinking as something under structural as well as biological control.

Finally, chapter 6 examines how researchers at Coast interacted with the popular press and evaluated media coverage of their work. Communicating in public presented researchers with a different set of conflicts than they faced in the laboratory—whereas in journal articles and grant proposals they could default to making cautious claims, when speaking to nonscientists they saw compelling reasons to make stronger statements. It was at precisely this moment when calibrating claims was especially important that they simultaneously experienced the breakdown of their most effective tools for claims management. Examining communicative practices at the boundaries of the laboratory reveals the limits of practitioners' control over narratives about animal experiments and human disorders.

CHAPTER 1

Containing Complexities in the Animal Behavior Genetics Laboratory

Ordinarily after the introductory behavior genetics class at Coast University, the graduate students dispersed almost immediately, but today everyone was unusually still. We were in a small classroom deep in the bowels of the teaching hospital, and Dr. Laura Martin, a senior investigator in the Department of Neuroscience, had just finished delivering a lecture on environmental interactions and mouse behavior. In her typical direct manner, she had presented experiment after experiment showing the variety of ways in which the environment could change a mouse's behavior. Not only could experiences in the womb or parental behavior early in life impact a mouse's later behavior, but even the placement of the cages in the mouse housing rooms or the light levels in the testing space could alter behavioral test results. Each slide layered on a new set of variables affecting mouse behavior until I felt as though I'd been buried in an avalanche of competing factors. Considered individually, each one of these factors seemed reasonable enough to take into account when setting up behavioral experiments, but their collective impact was overwhelming. Judging by my fellow graduate students' contemplative faces, I was not the only one feeling a mix of awe and frustration by the end of the lecture. In the semidarkness of the classroom, illuminated by the glow from Dr. Martin's final PowerPoint slide, this moment of shared stillness suddenly felt quite intimate. "With all of this complexity," one of the aspiring scientists remarked, breaking the silence, "it's hard to feel like you have a prayer." Dr. Martin nodded and reassured them that they were all in the same boat. Sometimes, she said, she felt depressed by it, too.

"Complex" was one of the most frequently used adjectives I heard during my time at Coast. Mouse behavior was complex, according to Coast researchers, as were the experimental setups used to test it. The resulting data sets and their interpretation were complex as well. The genetic factors underlying these complex behaviors were themselves complex, full of multiple interacting factors. As the scientists talked, complexity often began as a quality of the entities that they dealt with and transformed into an independent entity of its own that they grappled with. Researchers talked about complexity as though it was the ghost in the machine animating the objects they dealt with, making genes and neurotransmitters and mice behave in unpredictable, inscrutable ways. The mood that scientists slipped into during such conversations resembled the mix of awe and frustration I had experienced in the introductory behavior genetics class. Researchers talked about the complexity of biological processes reverently but also with a barely contained sense of exasperation that behavioral phenomena continually overflowed the boundaries of the experiments that they had so carefully constructed to contain those complexities.

This chapter explores what researchers at Coast meant when they described things as complex, and how they arrived at a sense of what it means to describe behavior in this way through their laboratory work. Describing entities or processes as complex is increasingly commonplace in the physical and life sciences, but the term signals different things to different practitioners. In physics and computer science, practitioners often employ "complex" in a quite specific sense to describe emergent phenomena arising from the interactions between individual components in a system. Marking something as complex, in these fields, means that the phenomenon is one that must be understood at the system level. As the physicist Philip Warren Anderson (1972) put it in a widely discussed essay, this view of complexity holds that "more is different," in the sense that the properties of the aggregates of many objects cannot be captured by studying the individual objects in isolation—they must be studied as an interacting system. In the life sciences, the term is often used more loosely to describe systems with many components, which may or may not have emergent properties that can only be understood at the system level. This understanding blurs the boundary some practitioners might draw between the "complex" and the "complicated"—that which must be understood at the system level versus that which in principle could be reduced to the sum of the contributions of individual components.

Researchers at Coast used the term in a variety of ways, only some of which overlapped with the definition that a physicist or computer scientist might provide. When some researchers invoked complexity, they were expressing a commitment similar to Anderson's—that behaviors could not be studied reductionistically, one gene at a time. But for other researchers, complexity meant only that behaviors were multigenic and that there was no single "gene for" a particular disorder waiting to be discovered. In other situations where researchers invoked complexity, they seemed to be making a claim not about the nature of behavior at all but rather about the difficulties or frustrations they experienced in trying to study it.

In light of the polysemy of Coast researchers' uses of "complexity," I argue that their complexity talk is better understood as expressing epistemological commitments rather than ontological ones. As other analysts have argued, scientists' uses of the term serve functions other than making claims about their views on the nature of the phenomena they are studying. As Arribas-Ayllon, Bartlett, and Featherstone (2010) argue, describing behavioral disorders as "complex" performs rhetorical work for knowledge communities by accounting for past failures in their field. Complexity explains why previous research efforts might have produced inconsistent findings about how genes impact psychiatric disorders, and it constructs careful optimism about the promises of new methodologies and about what the field can hope to accomplish.

For the behavior geneticists at Coast, complexity talk served to cultivate shared stances on knowledge production while allowing for a certain degree of ontological heterogeneity within the community. Despite the ubiquitous use of complex as an adjective, researchers at Coast did not necessarily share a unified vision about the underlying reality of behavior. What they did share, however, was an agreement that working from the assumption that behaviors emerged from the interaction of multiple, small genetic and environmental factors was the best way to produce credible scientific knowledge about them. In describing behaviors as complex, researchers articulated assumptions about what kinds of barriers might lie between them and an understanding of behavior and how a good practitioner should attempt to chart the journey ahead.

This chapter outlines some of the epistemic problems researchers at Coast associated with complexity and how they attempted to deal with them. Researchers attempted to contain the complexity of psychiatric disorders through their experimental practice by breaking down behaviors into smaller units for analysis and creating controlled laboratory environments in which to study them. In practice, however, these measures were difficult to achieve. In the absence of the types of techniques and circumstances that researchers thought would allow them to generate conclusive statements about the genetics of behavior, the knowledge that they produced took on a permanently provisional quality. The experience of being socialized into this epistemic stance resulted in what I describe as a "complexity crisis" for new practitioners, where they began to doubt not their understanding of the nature of behavior but their ability to study it.

COMPLEX BEINGS, COMPLEX BEHAVIORS

"Step one: Don't anthropomorphize," instructed the opening paragraph of a textbook chapter on animal models of psychiatric diseases (Crawley 2000, 179). I was reading the introductory textbook, titled *What's Wrong with My Mouse?*, at a bar on campus in preparation for my first visit to the Smith Laboratory. After reading through many chapters outlining methods for assessing the general health, motor functions, and sensory abilities of laboratory mice, I had finally arrived at the section on animal models of psychiatric disorders. I took a sip of my beer and read on. Jacqueline Crawley (2000, 179), the author, continued: "Emotions are personal, internal, and highly species specific. There is no way for a human investigator to know whether a mouse is feeling afraid, anxious, depressed, or experiencing hallucinations. These are subjective emotional experiences, existing in the mind and body of the individual. Major mental illnesses involve neural circuitry that may be uniquely human. . . . Aberrant behaviors symptomatic of human mental illnesses, therefore, may not occur in a recognizable form in rodents." A leader of the field opening with caveats such as these, I thought, did not make for an especially auspicious introduction to the practice of animal modeling. When the second edition of the textbook was published in 2007, a few months before I started my longest stretch of fieldwork in the Smith Laboratory, Crawley's (2007, 227) opening message had been further intensified—the phrase "don't anthropomorphize" was now set in italics and punctuated with an exclamation mark.

Crawley's cautionary introduction highlights several of the difficulties animal behavior geneticists associated with using complex organisms to model complex human disorders. Researchers described the disorders that they dealt with as complex both because of the types of characteristics that made up clinical categories such as "alcoholism" or "anxiety" and because of how those characteristics were grouped together. Many core features of psychiatric disorders—such as persistent worrying or the cravings for alcohol or drugs that addicts report—are what researchers described as internal mental states. Finding ways to measure those traits in humans, who can discuss their experiences with researchers, was difficult enough. But figuring out how to identify and measure analogous traits in animals presented a whole new set of challenges. As Dr. Smith pointed out to me in one of our first meetings, the criteria outlined in the *Diagnostic and Statistical Manual of Mental Disorders* (DSM) for diagnosing alcohol dependence range from those that could be construed as external biological traits observable across species, such as evidence of withdrawal symptoms, to those that are much more subjective and specific to humans, such as giving up important social and occupational activities due to alcohol use. He told me that he thought that the majority of the criteria in the DSM-IV were things that were "pretty tough" for animal models to capture: "Five of those seven symptoms are things like losing your job or persistently continuing to drink even in the face of evidence that your health is falling apart. You know, the doctor telling you you're killing yourself, your liver's getting trashed and you keep drinking . . . your relationships fall apart, you wind up in jail, you're obsessed with getting the drug. Those are all very human symptoms that are at the core of the disorder." Even if researchers could devise ways to model the subjective emotional experiences central to psychiatric disorders, these lines of research would run up against long-standing debates about whether humans can make credible statements about the mental states of other species. The question of whether the animal mind is an appropriate topic for scientific investigation has been quite divisive in the history of the behavioral sciences, with entire fields built around the central premise that only animal behavior, and not the animal mind, is accessible to scientists (Crist 1999, Daston and Mitman 2005).

The "checklist" approach to diagnosis employed by the DSM and other measurement tools used in humans also drew criticism from animal researchers for how it lumped together a variety of different traits. A DSM-IV diagnosis of generalized anxiety disorder, for example, required that a patient show excessive anxiety for at least six months along with any three of a list of six symptoms, such as irritability, muscle tension, and sleep disturbance. The entry for "alcohol dependence" similarly had a list of seven symptoms, any three of which can form the basis for a diagnosis. The separate entry for "alcohol abuse" required that a person exhibit one or more symptoms from a list of eleven. What this meant, Dr. Smith joked to me, is that when the different possible combinations of symptoms were parsed out, there are "1,200 some odd ways you can be an alcoholic."[1]

In this sense, anxiety and alcoholism were complexes—amalgams of different clinical features, which may or may not share the same biological causes. To some extent, researchers treated this heterogeneity as an unavoidable problem of studying high-level behaviors that developed over time in response to multiple genetic and environmental inputs. They argued, however, that the imprecision of the diagnostic tools used to classify humans amplified this problem. While in the late 1990s and early 2000s researchers had speculated that genetic techniques might be able to carve out more specifically defined disorders from these broad diagnostic categories, a decade later this prospect seemed increasingly unlikely. Some practitioners advocated instead that researchers would be better off trying to define entirely new disease categories for study, ones that were more narrowly defined and more clearly based on biological characteristics. How basic scientists such as the animal behavior geneticists at Coast were to form research agendas around disease categories that were themselves contested and unstable was an open question.[2]

Using higher organisms (such as mice and rats) as models for human behavior introduced another set of complicating factors related to the individuality of the animals. Coast researchers seemed to have little difficulty talking about a colony of bacteria or a bottle of flies as analogous to other supplies of uniform chemical reagents in the laboratory, but they resisted talking about even the highly standardized laboratory mouse in the same way. They described other laboratory organisms—bacteria, viruses, worms, flies—as being productive because of their simplicity and uniformity, in line with historian Robert Kohler's (1994) description of the fly as a kind of laboratory technology that found great success because it could be mass-produced. But while researchers did talk about mice as laboratory tools, to some extent, they also described them as individuals. Researchers argued to me that their own scientific training and the complexity of the animals themselves prevented them from viewing mice as completely interchangeable with each other. As Dr. Scott Clark, a longtime collaborator of Dr. Smith and Dr. Martin's, explained, "Brains change, right? It's not the same thing when I test *this* mouse and *this* [other] mouse. This mouse is different! Even if they're genetically identical, they're different! This mouse may have come from a cage where it grew up with two siblings, and this one with four siblings . . . that['s the] nuanced level of complexity that psychologists would naturally be interested in." This was even truer of the rhesus monkeys used at the affiliated primate research center that many of the Coast students rotated through in their early years of graduate school. Even long after individual monkeys had been turned into data points, researchers recalled their quirks, proclivities, and personal histories. When Dr. Sherry Trudeau, the charismatic head of the neuroscience division at the Primate Center, talked about her research, she would often circle particular data points with her laser pointer and tell the audience the name of that monkey and a bit about its personality.

Researchers talked about the unique properties of higher laboratory animals as both a resource for and a barrier to knowledge production. On the one hand, researchers at Coast argued that the "complexity" of the "natural behavioral repertoires" of mice was useful for modeling psychiatric disorders, and they saw the variation in behaviors that mice displayed as truer to the human scenario. Dr. Clark, for example, told me at length about the burrow systems that some rodents will form, how they vary with social dynamics and population size, and how mice emerge from and retreat into these burrows in response to potential threats. He thought that these burrowing behaviors could be a useful model for human anxiety, which also varied in response to social dynamics and perceived threats in the environment.

But drawing on these variable natural behavioral repertoires also complicated the experimental scenario. Dr. Marcus Lam, a postdoc who had recently joined the Smith Laboratory after finishing his doctoral degree in a pharmacology program, told me that he was amazed to find that many of the Smith lab's common experimental protocols were designed in such a way that the mice in a particular study might not all get the same dose of alcohol. While allowing mice to drink from a bottle of alcohol might be a better reflection of the way that humans ingest alcohol, Dr. Lam said that he would rather just inject the mice directly with a fixed dose to avoid introducing "noise" into his data right from the very beginning. The same individual variation that made working with mice seem truer to the human scenario could also be viewed as grounds for questioning the quality of the scientific data.[3]

Setting one unstable entity—a lively and unpredictable mouse—in relation to another unstable entity—a loosely bundled amalgam of human traits—therefore created numerous possibilities for uncertainties, reversals, and interpretational difficulties in animal behavior genetics research. Even though both human psychiatric disorders and laboratory mice were entities that had been the target of much effort to standardize, characterize, and contain them, they were also entities that continued to overflow the boundaries of these categories.

Dr. Smith recounted a story to me that illustrates why researchers believed these complexities made the production of genetic knowledge about behavior extremely difficult. One of the founders of the field, Gerald McClearn, designed a project early in his career to look at the genetics of alcohol withdrawal in mice. He took a group of genetically diverse mice, measured them on twenty-one different behavioral tests related to alcohol withdrawal, and then mated mice with high scores on this test panel to each other. Researchers had successfully used this selective breeding approach to study the genetics of other behaviors, such as Robert Tryon's now well-known breeding experiments to generate so-called "maze-dull" and "maze-bright" rats in the 1940s. In this case, however, it seemed not to work. After several generations of mating the high scorers and the low scorers to each other, the difference between the two groups was slight, and McClearn eventually abandoned the project. The failure of this project is all the more notable because some of the tests he was using measure traits that are now believed to be highly heritable, such as the tendency for mice to experience seizures while they are withdrawing from alcohol. Dr. Smith speculated that the project had foundered because McClearn's attempt to account for multiple facets of alcohol withdrawal in his panel of tests may have actually hindered his progress. Some of the mice in his study may have received high scores for withdrawal because they experienced seizures, while others might have scored high due to increased anxiety. Consequently, mating mice that were experiencing severe withdrawal for different reasons simply reshuffled the genetic deck in each generation rather than concentrating genetically similar individuals.

While one might conclude from this quick overview that using mice to model human behavioral disorders is simply more trouble than it is worth, animal researchers saw human research as equally, if not more, problematic. Watching animal researchers evaluate human studies brought this into sharp relief. In one journal club meeting I attended, a postdoc in the department explained that lately, he had come to the realization that he knew relatively little about alcoholism in humans compared to what he knew about animal alcohol models. And so, he had selected a paper for discussion on the relationship between anxiety disorders and alcohol use disorders in the Dutch population. The audience of animal researchers seemed unimpressed by the study. They quickly pointed out numerous "flaws" with the research design, from inconsistencies in data sets brought together to conduct the study to a reliance on suspicious self-reported data about the prevalence of those disorders. Later, Dr. Lam told me that reading the paper confirmed for him that he could never study human populations because they were far too "messy." Hannah, a graduate student in Dr. Martin's laboratory, concurred. She commented that human studies were not very satisfying to her, because "the proverbial grain of salt is just a lot bigger in human genetics."

BREAKING DOWN COMPLEX PHENOTYPES

One of the reasons that researchers at Coast found animal research to be a more satisfying way of producing knowledge was that animal behavior could be manipulated and segmented in ways that would be impossible in humans.

Human addicts, for example, might be using multiple drugs, or dealing with depression and anxiety in addition to substance abuse. Researchers studying animal populations, on the other hand, could examine particular drugs or phases of the life cycle of addiction in isolation.

Decomposing disorders such as alcoholism into smaller subunits held an intuitive appeal for those in the research community who subscribed to the view that behaviors were complexes, that is, aggregates of different biological phenomena. A version of this view was articulated in a highly cited paper published in 2003, where psychiatric geneticists Irving Gottesman and Todd Gould (2003) argued that researchers should focus on "endophenotypes" rather than behavioral "phenotypes." They envisioned endophenotypes as midpoints between genes and behaviors, units of analysis that were smaller than psychiatric disease categories and that might be more easily linked to particular genes. In their own work on schizophrenia, they demonstrated that characteristics such as dysfunction in eye tracking or working memory were common in schizophrenic patients and showed stronger links with particular regions of the genome than the larger category of schizophrenia. By adopting these kinds of units of analysis for their studies, Gottesman and Gould hoped that both human and animal researchers could accelerate their progress toward understanding the genetics of psychiatric disorders. Animal researchers, for example, could build their models around individual endophenotypes rather than attempting to model multiple features of a psychiatric disorder.

Gottesman and Gould's paper was first and foremost a methodological proposal, but it hinted toward an ontological project as well. They suggested that endophenotype-based analysis could help reorganize existing psychiatric disease categories and "establish a biological underpinning for diagnosis and classification" (Gottesman and Gould 2003, 641). While other practitioners also expressed hope about the capacity of biomedical research to break down diseases into more fundamental units, the belief that complex disorders were in fact composed of simpler, biologically grounded units that could be separated out using genetic analysis was far from a central tenant of behavior genetics research. For example, the prominent psychiatric geneticist Kenneth Kendler (2006, 1145) argued in a review article that while the idea was conceptually appealing, upon surveying the empirical evidence for these claims, he arrived at a "largely skeptical conclusion." It seemed unlikely to him that psychiatric genetics would "allow [researchers] to 'carve nature at its joints,'" and that "the project to ground [their] messy psychiatric categories in genes—as an archetypal natural kind—may be in fundamental trouble" (Kendler 2006, 1138).

However, breaking down psychiatric diseases into smaller units of analysis was a central component of most animal behavior geneticists' research strategy. Even if researchers were not confident that these redefined categories represented natural kinds, they could still form an important part of their approach to producing knowledge about behavior. For Dr. Smith, breaking down "alcoholism" into discrete phases (such as initiation, tolerance, and withdrawal) may not have been true to the complexity of the disorder, but it was a way to create a viable experimental program. He explained:

> So if you try to model these complex things, most of us start by being reductionists, whether we want to or not. We start as obligate, intentional reductionists. I'll say, OK, one of the symptoms of alcoholism in humans is tolerance. If you started out at twenty years old being able to drink three beers before you fell over and passed out, by the time you've been doing that for seven or eight years, your dose has escalated like crazy and now you can drink six beers before you fall over and pass out, and that's one definition of tolerance. You can model that quite easily in animals in lots of ways, and so there's been a lot of research on trying to understand what changes in your brain as you become tolerant.

For those who subscribed to the view that behaviors are complicated phenomena, breaking down these larger categories into their constituent parts fit comfortably with how they understood behavior. But for others, who took behavior to be composed of interacting factors that could not be cleanly separated, developing partial models was a trade-off. The reductionist approach that Dr. Smith advocated was not grounded in a belief that alcoholism was in reality comprised of simpler or more fundamental units; rather, it was motivated by the belief that this was a viable approach to generating some knowledge about it. Reductionism, here, was an experimental strategy rather than an ontological commitment.

Dr. Raymond Williams, a senior executive at the National Institute for Alcoholism and Alcohol Abuse (NIAAA), shared Dr. Smith's pragmatic view. He explained it to me as follows: "There's no one animal model for alcoholism. If you did, you would have drunk rats, but you don't. They don't like the taste of alcohol, you have to initiate them and get them to drink, right? So we're clear on that. It took people from outside the field a lot to understand that, that the animal models model one facet of the continuum that you see in alcohol dependence." Dr. Williams's description aligns with Jamie Lewis and colleagues' (2013) description of how scientific practitioners attempt to strike a balance between practical considerations such as keeping mice alive or obtaining ethical consent to work with them, and the scientific questions they want to ask. Even in cases where researchers are modeling single-gene disorders, Lewis and colleagues argue that researchers typically do not aim to recapitulate all features of the disorder but instead aim for a model that is "good enough" for particular practical purposes. The complexity of alcoholism and mouse behavior imposed practical constraints on alcohol researchers—rodents simply will not drink enough, for example, to reflect the intake levels seen in human alcoholics. Breaking down the work of generating knowledge about alcoholism into many models, each of which are "good enough" for representing a part of the disorder, was a pragmatic way of moving forward.

This experimental strategy may seem out of keeping with current trends toward "big data" and associated bioinformatics techniques in the life sciences. In the field of metabolomics, Nadine Levin (2014) argues that researchers have come to view multivariate statistics as the natural and correct way of engaging with complex biological processes. For metabolomics researchers, measuring hundreds of variables at once rather than studying individual variables in isolation is a way of engaging with the biological world that Levin (2014, 558) argues both "depend[s] on and reproduce[s] the notion that metabolism is complex." For researchers at Coast, multivariate statistics and tools that captured large amounts of genomic data were an important part of the arsenal for studying behavior, but they did not fully resolve the problems of how to engage complex phenomena. These tools, in their view, presented new interpretational problems that were just as challenging as the problems posed by attempts to decompose behavioral phenomena into smaller units. Dr. Martin described to me one ambitious study that her laboratory undertook to study the relationships between genotypes, drugs, and behaviors. This was a study, she said, that was unwieldy to execute but turned to be even more unwieldy to analyze. She recounted:

> One of my techs, many years ago now, ran a study that had fifteen inbred strains, three drugs, three doses of each drug and ten behavioral phenotypes, OK? That's as big as maybe a microarray almost, and it's a behavioral study! . . . I have pulled those data and tried to write that up multiple times, and I just can't seem to decide what the best way to put all of that together is. And that's pretty sad, actually! [laughter] It's a great data set that could tell us something about genetic commonalities among these drugs, and I can't decide, should I put this drug by itself, or should I put them all together? But when I try to put them all together, it becomes this

> monstrous thing. It's just an example of how, how do we handle this? And we really need better tools for that.

She joked that her technician would "like to come in and curse at [her]" for not publishing the study. But she felt she still lacked appropriate techniques for grappling with the large data set they had generated, and so it remained untouched in her filing cabinet. Breaking down psychiatric disorders into smaller units, in contrast, was a methodological strategy that felt familiar to many animal behavior geneticists, and it fit comfortably with a range of understandings about the nature of behavior. It could be a way to make visible the supposedly more fundamental biological kinds that some hoped to find underneath messy psychiatric categories, but it could also be a utilitarian strategy that allowed researchers to move forward in the absence of better tools for grappling with irreducibly complex behaviors.

This strategy, however, was a highly imperfect one. Dr. Smith described a project to me that he had been involved with, which was designed to group different animal models of "intoxication" into the more the specific skills those tests measured. A group of researchers took ten of the most popular tests of intoxication and sorted them into subcategories such as balance, muscle coordination, and cognitive abilities. They then designed a set of experiments to test the robustness of their provisional classifications, which quickly revealed numerous problems. Dr. Smith described one of the tests they examined, called the "grip strength test," where mice use their forepaws to grab onto a small bar attached to a force meter that measures how hard they can pull:

> So, you do that, you measure that, and alcohol makes them weaker, OK? And then we add another task where you have a window screen in a frame, and you take the mouse and inject it with alcohol, put it on the window screen and rotate it so it's horizontal like this [demonstrating with his clip board], and wait and see how long it takes before it falls off. With saline injection they can run around up there forever, but alcohol makes them fall off. No relationship whatsoever between those two tasks. So our ability to guess what each of those tasks was measuring was about nil.

Both the grip strength test and the screen test appeared to measure similar aspects of intoxication in mice that had to do with a mouse's ability to hold on to the force meter bar or the mesh on the window screen. But experiments revealed no relationship between these two tests: strains of mice that held on to the window screen for a long time while intoxicated did not necessarily do well on the grip strength and vice versa. The traits that researchers had grouped together under the supposedly more specific label of "grip strength" refused to cohere together in their experimental work, leaving the larger question about how specific factors contributed to the larger category of "intoxication" untouched. Researchers hoped alternately that they would someday find better classification systems or develop better techniques that would obviate the need for reductionism, but such methodologies seemed presently unavailable.

CONTROLLING THE EXPERIMENTAL SETTING

A second reason why researchers at Coast found animal research to be a satisfying way of producing knowledge about psychiatric disorders was that the environments of experiment animals could be carefully controlled. Unlike human populations, who varied in their diet or early life experiences even in controlled settings such as clinical trials, the lives of research animals could be managed in exquisite detail. The time and attention devoted to environmental control was a notable feature of experimental life at Coast. In my interactions with researchers, they often described themselves as "anal," "picky," or "perfectionistic" people, and they expressed pride in the amount of care they took in executing experiments and the lengths they were willing to go to ensure consistency. Chloe, a graduate student in the Martin Laboratory, commented admiringly on a research paper that we read for the introductory behavior genetics class, exclaiming to me that "they controlled for things that I never even thought of!" When I interviewed fellow grad student Hannah, I asked her if she could think of any controls that people in her lab used and that were probably unnecessary. She replied:

> Probably not, because I think I'm more obsessive than the others in my lab. I can tell you one thing that they do that I don't like is when they're doing these holding cages, a lot of times as they're setting up, they'll put a home cage on top of a holding cage so they can get that next group of mice out.[4] And there's, you know, a mouse in there. Well, that's a different experience for that mouse than for the rest of the mice, because it's dark, something's on top of it, you've got something new to explore, it's more noise, it's more shaking. So I just sit there and just grimace whenever I see that.

What was especially amusing about Hannah's complaints about her fellow lab mates was that as I went on to interview more members of the Martin Labora-

FIGURE 2. Handmade sign warning passersby about the impact of everyday noise on the animals, posted on the door of a procedure room at Coast University. Photograph by the author, May 2008.

tory, almost without exception every person described him- or herself as the most perfectionistic person in the lab and shared similar stories of others' questionable experimental practices.

In contrast to molecular biology, where the purity of the reagents, precision in measurement, and timing of reactions are the subject of intense concern (Jordan and Lynch 1992; Stevens 2011), it was the sensory aspects of the laboratory environment—the sounds, light, and smells in the spaces they worked—that attracted the most scrutiny at Coast (figure 2). While researchers generally treated the mouse genome as an entity that was controllable and standardizable, they portrayed the laboratory environment as much more unruly. When I asked them to list the five most important things that they controlled for in their experiments, they tended to list a few key biological characteristics of their test subjects (such as the genotype, age, and sex of the mice) and a wide variety of features from the housing and testing environment.

When I posed this question to Dr. Smith, he produced without hesitation a long list of environmental factors that he would expect to see listed in any behavioral paper, regardless of the specific type of experiment being performed: the size of the cage and the number of mice kept in each cage; the kind of ventilation system provided for the cages; food and water availability; food and bedding type; the temperature of the mouse rooms; the light/dark schedule in the rooms; and whether experiments were done during the light period or the dark. The building-wide shift to daylight savings time was announced in the Smith Laboratory's weekly meetings no less than a month in advance (and every week thereafter) so that researchers could prepare to incrementally shift their mice to the new schedule. Similar effort was directed at managing the possible effects of the animal housing facility's decision to switch over to a new brand of bedding made from recycled paper instead of corncob. One professor in the department was convinced that the specific type of plastic used to make the mouse cages had an important effect on mouse behavior. Dr. Smith was skeptical of this claim but told me that he always made a note about which cages he used in which experiments—just in case. Noisy events such as construction work, clanging carts, or people talking on cell phones in the hallway were also worrisome to Coast researchers. The upbeat music that was a familiar part of the molecular biology laboratories I visited was notably absent, and even the technicians gossiped in hushed tones while they were running experiments.

Researchers at Coast took measures to control these sources of environmental variation that often seemed quite extreme to me as an outside observer. Following the announcement of the animal housing facility's impending switch in bedding brands, more than one researcher I knew started accumulating secret stockpiles of bags of corncob bedding to use in the cages of their especially valuable mice. Matthew, a graduate student at Coast, was worried about what he called the "weekend effect." He pointed out that the environment of the research facility is different during the week, when many people are around, and so he avoided coming in on the weekend to run behavioral experiments, even though this fit better with his schedule. Jeffrey, one of the Smith Laboratory managers, described how he had resorted to holding a giant pillow over the fire bell during past drills to minimize the noise when the building managers could not be talked out of testing the alarm system on the floor that housed the mice. Alex, another graduate student in the department, was particularly concerned about his own smell. In an interview he told me that he once forgot to wear deodorant on the first day of a two-week-long experiment and he did not wear deodorant for the rest of his experimental run so that the rats would not sense a change in his body odor. When I recounted this story to other graduate students, to my surprise they agreed that this was a reasonable precaution. Another graduate student admitted to me that she had used the same brand of shampoo and conditioner since starting her doctoral research at Coast for this exact reason. The laboratory was not only a space where researchers could exercise control over the environments of the animals, but one where they also had to exercise control over themselves.

Coast researchers' inclusion of themselves as part of the laboratory environment reflected an understanding of animal behavior as a dynamic, relational entity. Robert Kirk (2009, 2014) traces the roots of this view of behavior to intersections between ethological thinking and pharmacology in the 1950s, such as British researcher Michael Robin Alexander Chance's research investigating how social behavior in animals could alter physiological reaction to drugs. It was at the intersection of those fields, Kirk (2014, 249) argues, that researchers began to situate "the knowing subject within the environment of, and therefore in relation to, the object of (or means to) knowledge." He argues further that this relational framing encouraged reflexivity and explicit consideration of methodological problems in animal behavioral research. If the researcher was not simply a passive observer but an active agent in the experimental scene, then researchers needed to attend to their own behavior if they wanted to produce reliable experimental knowledge.

When it came to behavioral experiments, Dr. Smith and his colleagues seemed willing to entertain the possibility that almost any change to the laboratory environment could potentially change their results, making the creation of a truly controlled laboratory environment seem a nearly impossible task. Dr. Smith readily admitted that controlling all of the variation that might impact mouse behavior was simply infeasible, and so he himself adopted a pragmatic approach: "I mean maybe humidity is important, you know? I don't even measure the humidity in my lab. I can guess what it is from [Coast City's] weather records, but you know, I'm not going to go there. Partly that's a practical thing. You know, so you find that it's humidity. So what? What are you doing to do next? [laughs] People are not going to start changing the humidity in their labs."

Meanwhile, at a university in the Eastern United States, Dr. Charles Westin (a former member of the Smith Laboratory) was worrying about precisely this issue while setting up his own laboratory. He had been offered space in a centralized animal housing facility, but the facility did not offer as much control over the environment as he would have liked, and humidity was one of the factors that "bugged him." During the summer, he knew that the humidity was increasing in the mouse rooms, but there was little to be done about it. He told me he had thought about bringing a dehumidifier into the facility, but decided against it because the noise of the dehumidifier would introduce yet another source of environmental variation.

Environmental controls, and failures to maintain control, took on a particular sense of urgency in light of researchers' assumptions about complexity. As one animal behavior geneticist put it in a widely cited paper arguing for the importance of a particular methodological control: "It must be remembered that behavioral and neurobiological traits are fairly complex, often variable, and are most probably influenced by a large number of genes as well as environmental factors. In order to dissect such traits and to understand the complex web of interactions among the underlying biological mechanisms, it is crucial to provide adequate controls for as many variables as possible" (Gerlai 1996, 180–81). If the genetics of psychiatric disorders were straightforward, with one or two genes accounting for much of the disease phenotype, then minor variations in the experimental system mattered less—the effects of those genes were likely to be visible even in the presence of noise from the laboratory environment. But if behaviors were complex, composed of multiple and possibly interacting factors, then researchers had to be able to create highly controlled settings in order to produce knowledge about them.

Once again, this methodological strategy fit comfortably with a range of different understandings of complexity. For practitioners who understood behavior to be a system-level phenomenon, taking into account the fullest possible range of environmental and genetic factors was crucial for understanding the conditions under which behaviors emerged. For those who saw behaviors as the sum of multiple genetic and environmental factors, exercising control was simply a prudent measure in light of the present lack of knowledge about which environmental factors mattered for which behaviors. When I interviewed Dr. Martin about the control measures her laboratory had instituted around animal housing, I asked her why she was so cautious about using only mice bred in their own facility when one of the articles she had assigned for the introductory behavior genetics class had examined precisely this issue and found no behavioral differences between mice raised on site versus mice shipped from a different facility. She explained:

LM: That's right, for the traits that they were looking at. But you can't be sure for every trait, right? So that's another thing that we always look at, too. In a given experiment, we try very hard to use only animals that are born here, or only animals that are shipped so that we're controlling that, just in case for that particular trait it matters, because you know for all of the traits that they looked at it didn't matter, but who knows? It might matter for something else.

NCN: It could be different for a different trait.

LM: Yeah.

NCN: Or could it be that that's something that everyone thinks is really important, because it makes sense, but it actually turns out that it's not?

LM: It may not be, but you might as well control it, just in case.

Dr. Martin asserted that because researchers did not yet know all of the variables that were relevant for a given behavior, controlling the experimental environment was a safe strategy. If a particular psychiatric disorder turned out to have a more straightforward genetic etiology, no harm would come of exercising strict controls.

But again, as in the case of breaking down complex phenotypes into smaller units, this was an epistemic strategy that was difficult to execute in practice. The daily disruptions that are a part of life in any laboratory—broken equipment, changes in staff, delayed schedules—contributed to researchers' impression that the laboratory environment produced a virtually endless supply of unaccounted-for variation. When combined with the assumption that even small changes to the experimental situation could alter complex behaviors, the laboratory environment became a considerable source of angst. The practical realities of laboratory life made the conditions for generating good knowledge about the genetics of behavior seem to be just out of reach.

REDRAWING THE BOUNDARIES OF EPISTEMIC CONCERNS

The role that complexity talk played in the laboratories at Coast, then, was not to secure agreement on the likely composition of the molecular underpinnings of psychiatric disorders but to direct attention to particular epistemic problems. The broad assumption that behavior involved multiple genetic and environmental inputs, and that it developed over time in response to those inputs, translated into a call for a specific set of experimental habits and sensibilities. Complexity talk served to redraw the boundaries of what researchers could ignore and what they needed to attend to, thereby readjusting researchers' expectations about their own role in shaping the phenomena they were studying. Practices designed to contain complexities encouraged researchers to view themselves and their laboratories as part of the set of factors that produced behavior, which transformed everyday problems into matters of intense epistemic concern. If behavior was not a stable entity waiting to be made visible through experiment but rather a phenomenon that researchers themselves helped to constitute, then changes in researchers' practices or the laboratory environment were potentially epistemically significant.

The specific ways in which assumptions of complexity shaped Coast researchers' practice was especially apparent in how they reacted to unexpected results in different kinds of experiments. When behavioral test results departed from the norm, researchers often described those incidents as manifestations of complexity. Graduate student Matthew, for example, described an experience where a behavioral test he was running produced an unexpected result in the following way: "The more complex behavior is, the harder it is to get it to work. And so for whatever unknown reason, things don't always turn out as you expect them to. . . . And so there are just difficulties with that, to get it set up and running. It's not so much a technical problem, it's just something that since it's so complicated, you can't figure out what's going on that you're not controlling for." The experiment that Matthew was describing was a "conditioned place preference" test, a protocol that researchers use to measure an animal's preference for a drug by teaching them to associate a distinctive space with receiving the drug. Prior to this training process, mice are not supposed to exhibit any particular inclination for one side of the apparatus or the other; yet, Matthew's mice unexpectedly spent most of their time in one space. Matthew explained to me that this particular test measured an especially complex phenomenon that depended on multiple interlocking behaviors and capacities, such as learning, memory, and spatial orientation. Even though he had been as careful as he could be about controlling the conditions of the experiment, he still expected that the test might produce something unexpected, and attributed the unexpected result to as-yet-unidentified features of the environment he had not controlled for. He explicitly marked these unexpected results out as something that was "not a technical problem" or something attributable to his own lack of skill as a researcher, but rather due to the difficulties of working with complex behaviors.

There were many other instances, however, in which researchers at Coast treated unexpected results as simple malfunctions. Complex phenomena could generate surprises, but straightforward or established phenomena generated only technical aggravations. Dr. Tremblay's laboratory, for example, devoted much attention to troubleshooting a protocol for detecting one of the brain proteins they were interested in during the months that I stayed with them. The protocol had repeatedly failed to generate the expected results, and the Tremblay Lab was engaged in a process of scrutinizing and manipulating each component of the offending protocol, such as the exact time a reagent was added to the mixture, or how many times the antibody for the Western blot was thawed and refrozen. But unlike researchers' descriptions of behavioral test malfunctions, there was never any question about whether these methodological difficulties interacted with the proteins being studied. Dr. Tremblay and her students regarded brain proteins as stable entities that were insensitive to the assays they used to interact with them. A blank gel was not an occasion for reflecting on the nature of the protein or researchers' means of knowing it; it was a signal that the test was still malfunctioning and had failed to make the existing proteins visible.

It is tempting to describe these differences in the way that researchers dealt with unexpected results in terms of how close the tests and phenomena in question were to the "warm unsettled part of the research front," as Latour and Woolgar (1986, 96) put it. To use Hans-Jörg Rheinberger's (1997) terminology, we could say that the first scenario dealt with an "epistemic thing," whose contours and responses to manipulation were still under investigation, while the second dealt with a "technical thing," which researchers assumed they understood well and were using as part of the infrastructure to set up new experiments. What makes this distinction difficult to hold on to in analyzing research work at Coast, though, was the long expanse of unsettled research questions where epistemic things seemed to bleed into technical things. The conditioned place preference test is a protocol that has been used in the field for decades, and it was the well-characterized initial baseline behavior that went awry in Matthew's experiment, not the results from an experimental manipulation. The long history of use of this test and the fact that it was being deployed in the service of investigating other experimental questions did not seem to prevent Matthew from interpreting the anomalous results he generated as a manifestation of behavioral complexity rather than a malfunction or his own error. What allowed even established experimental systems to act as "generators of surprises" (Rheinberger 1997) was Coast researchers' assumptions about the emergent nature of behavior. Given a different set of orientations and assumptions, it seems entirely plausible that a researcher could view the abnormally behaving mice in Matthew's experiment as akin to a bad batch of a chemical reagent. But because Matthew regarded behavior as something produced through the process of investigating it, how he smelled or where he placed the testing chamber in the room could therefore change the behavior itself, not just what kind of measurement his test produced.

If this sounds like a relatively unsurprising insight, it is perhaps because this view of scientific practice resembles ideas that are foundational to STS. The researchers at Coast may not be willing to go as far as Karen Barad (2003,

816; 2007) in her redefinition of the experimental apparatus as a "dynamic (re)configuring of the world," but they operated with a similar sense of the indeterminacy of the apparatus's boundaries. Barad's (2007) description of how an experimental physicist's cheap cigar and the sulfurous smoke it produced was central to the outcome of his experiment would certainly resonate with their experiences—indeed, whether those who came into contact with the animals were smokers was one of the many parameters they tracked.[5] Their understanding of behavior as emergent also shares some commonalities with Annemarie Mol's (2002) argument that actors bring multiple ontologies into being through their practices. Just as Mol argues that primary care providers, surgeons, and patients each "enact" atherosclerosis differently, researchers at Coast believed that each practitioner and each behavioral test brought a slightly different manifestation of "anxiety" or "addiction" into existence. Researchers at Coast even placed some of their beliefs within the boundaries of the experimental system, pointing to psychologist Robert Rosenthal's work on experimenter expectancy effects as evidence that what a researcher believed could alter an animal's behavior.[6] Of course, there are many points on which Coast researchers and STS scholars would not be aligned. Researchers at Coast unsurprisingly talked about psychiatric disorders in a realist mode, asserting that there were true and false statements that could be made about them. But even though researchers at Coast may not have understood psychiatric disorders to be as actively produced as STS analysts might, they did not treat them as stable entities out in the world, waiting to be discovered. In some instances, this was because researchers thought the perfectly controlled environments or neatly delineated categories that would allow them to treat behaviors as fixed objects were presently unachievable; in other cases, they believed behaviors to be fundamentally different in kind from other scientific objects. In the near term, these different positions converged around an agreement to treat behavior as a lively and unstable entity, and the knowledge researchers produced about it as provisional.

A CRISIS OF COMPLEXITY

In one sense, the types of problems and practices that I have described here are common to all experimental work. Coast researchers are hardly alone in portraying meticulousness as an admirable quality in an experimenter or in emphasizing the scientific importance of experimental controls; and as much STS research has shown, natural phenomena do not present themselves in neat packages, and what happens in the laboratory is the laborious and messy work of transforming amorphous entities and recalcitrant objects into crisply delineated things that can appear to stand on their own. Accomplishing these transformations involves many surprises, reversals, and frustrations, which make the daily work of the laboratory seem quite unlike the planned and rational process described by the conventional scientific method. Rheinberger (1997) has described experimentation as the process of negotiating a labyrinth that is in the process of construction. The existing walls of the maze both limit and orient the direction of the new walls to be added, and these walls both blind and guide the experimentalist as she moves through the maze. "A labyrinth that deserves the name is not planned and thus cannot be conquered by following a plan," Rheinberger (1997, 74) writes. "It forces us to move around by means and by virtue of checking out, of groping, of tâtonnement."

But even this elegant construction metaphor doesn't capture the full extent of the instability in the experimental terrain that Coast researchers described. Despite their best efforts to contain the complexities of behavior—to shape the phenomena, their experimental techniques, and themselves to make behavior tractable—the knowledge they produced remained unstable. If the experimental system can be thought of as a labyrinth, then researchers at Coast experienced them as labyrinths whose walls might suddenly shift behind them, obscuring their view of where they had come from as well as where they were going. Recently produced results retained the lingering uncertainties associated with their circumstances of production, and even supposedly well-established behavioral phenomena could suddenly disappear. Graduate students felt this instability especially keenly, because their status as trainees meant that their expectations about experimental practice were still in the process of being formed and the stable interpretations they had held as undergraduates were being actively unsettled. In interviews where I asked students about their trajectory through the program at Coast, many described how the training they received made them reexamine their prior experiences in the laboratory. Liam, a graduate student who had just finished his first-year rotations through different laboratories in the department, told me that this was a formative experience that shaped the way he thought about experimental work:

LIAM: I guess in trying to set up all of those experiments, I started to learn a lot about well, geez, all of these things really factor into the animal's behavior and you really need to control for all of it. So I guess that's where I really started to learn about that.

NCN: "All these things," like you mean things from the lab environment?

LIAM: Things from the lab environment, all the different things I needed to control for, that's kind of a vague word, sorry. I don't know. Everything that's in the animal's environment, the injections, handling the mice, you know, I didn't really think about the handler effects, and so as an undergrad, I did research with a bunch of different people, and so we would all sort of take turns running some of the experiments so that we all weren't in there all the time. But I didn't really realize that maybe that's actually affecting our results because the animals are getting exposed to different things because of the handler effects. And so when I came here, I really learned how important all of these controls were, and how important it was to control all that.

In light of these new assumptions, prior experiences often took on a different character. When I asked graduate students about their research experience prior to coming to Coast University, many told me about projects that they had completed and had thought were successful, but that they now realized were in some way "bad projects." Hannah, a student in the Martin Laboratory, described an experiment that she had designed for her senior thesis where she looked at whether a drug that blocked a brain receptor could alter a rat's motivation to drink alcohol. She tested rats on a simple maze where one side had a bottle of alcohol and the other side was empty. After training rats on this maze, she gave her rats the drug and measured whether they went toward the arm with the bottle of alcohol or the empty arm of the maze. She told me that she realized now that this experimental design was "wrong," because she didn't control for other possible effects of administering this drug on her rats' behavior. At the time, she assumed that her experiment worked and the drugged rats were slower in seeking out the alcohol bottle because they were not as motivated to drink. In looking back at her old experiment, now Hannah said that she "just didn't know enough of the caveats," such as the possibility that the drug she gave her rats might have impaired their movement. The stories I will explore in chapter 2, where senior faculty at Coast recounted moments where their own seemingly solid experimental findings had shifted in front of their eyes, further contributed to this impression of instability.

For many of the graduate students, the process of undoing what they had previously taken as constant and replacing those beliefs with a series of interacting factors to be managed was psychologically stressful. Many recalled that they went through what could be described as a "complexity crisis" at some point in their graduate training. They described moments where, like the new graduate students in the introductory behavior genetics class I described in the opening of this chapter, they felt overwhelmed by all of the new factors that they had to account for in their experimental practice. After being inundated with information about all of the factors that could alter behavioral experiments, students reached an impasse where they questioned whether there actually was a path forward that would allow them to extract stable scientific information.[7] Even after breaking down complex phenomena into smaller units for study and going to extreme lengths to control the laboratory environment, behavioral phenomena still sometimes manifested themselves in the laboratory in baffling ways.

Emily, a senior graduate student in Dr. Ruth Martin's lab, described working through this sense of defeat as a rite of passage that all students went through at some point in their graduate training. In a conversation we had about her ongoing dissertation work, she told me about how her laboratory had identified a promising region of the mouse genome that seemed to influence alcohol withdrawal. Their attempts to pinpoint a specific gene, however, had not been going so well. There were too many genes in the region to say with certainty which one might be creating the effect, and when they inserted one of the genes into another mouse strain, the results were inconclusive. They were not sure if they had been "misled" by the gene or if there was a problem with the animal model. Consequently, Dr. Martin had suggested that they go back to the original data that had directed them toward that candidate gene—which, on reanalysis, revealed what she described as a depressingly large number of alternative candidate genes. As Emily spoke, I began to feel that increasingly familiar, heavy sensation of being weighed down by so many factors to consider that even small steps toward clarity seemed as though they would require enormous effort. These reactions must have registered in my expression, because Emily quickly redirected the conversation to reassure me: "Ruth and I joke about this all the time, because it's like every question answered presents a million more questions, and we're like, well, it's job security. When you start doing this kind of work with complex traits—it's like you realize at some point what exactly you're getting into, and you're either going to run the opposite direction because you're never going to find an answer, or you just think—I don't know, you have to make peace with it and be like I'm going to do what I can do to help."

She explained to me that the kinds of problems she had been describing were the things that behavior geneticists had to expect they would come up against because of the phenomena they chose to study. If students could not find a way of coping with the complexity, then they often quit or transferred to a different field. But, admitting to me that remaining optimistic under such circumstances was easier said than done, Emily also shared with me one of her favorite coping mechanisms: when she started to feel hopeless about her potential for making progress, she watched episodes of the reality TV show *Intervention*, where addicts are confronted by their families and offered the chance to go to a treatment facility. Watching stories about people with addiction problems and the devastating effects that the condition had on their lives recommitted Emily to moving forward with her research, even when the progress felt agonizingly slow.

Hannah, who was also a senior student in the program, had a similar but slightly less optimistic take on coping with complexity. She told me that even though she had largely made peace with the difficult reality of studying behavior, she still had moments where she felt overwhelmed by the task at hand: "I still feel like I need to be like, well, we really don't know, because it's all [pauses]—which I think is really just a common thing with students, seeing this complex thing where it's like, oh my God, there's nine hundred things that affect this! And those things affect each other, which then affects that, which affects this. So the whole thing is just all over the map, and you're just—confused, wondering how in the world you're supposed to say anything." Her coping mechanism, she told me, was to cultivate a kind of tunnel vision that would allow her to focus on the two interacting brain systems she had chosen to study for her dissertation. And yet, she felt continually uneasy about the other connected factors that she was purposefully ignoring. "There's so much more to know," she told me. "But you can't. You can't."

PRODUCING A "TRAIL OF INTRIGUING HINTS"

Those who came through this trial by complexity emerged with revised expectations about what kinds of scientific statements they would be able to make. The types of facts that researchers aimed to generate were partial and provisional, ones that they expected would vary between species and laboratories, and ones that they hoped would be replaced by more stable findings in time. Dr. Scott Clark, Dr. Smith's collaborator, seemed almost calloused to the surprises and reversals that researchers associated with complexity, and he was very cautious about potential research outcomes. One experience he described to me illustrates how dealing with "complexity" can alter researchers' stances on knowledge production. He was working with two strains of mice that had been selectively bred using the same procedure, and testing them with a series of drugs that either activated or blocked a brain receptor he was interested in. When he got to one drug, he found that it had exactly the opposite effect in the two different strains. He had predicted that the drug in question would "sober them up a bit," and while that prediction held true in one of the mouse lines, it made mice from the other line more drunk. When his graduate student first showed him the data, he was sure that she had done something wrong, and he asked her to repeat the experiment. She repeated the entire experiment six times before they eventually published it. But Dr. Clark told me that even though he had seen the surprising effect enough to "believe" it, he still could not explain it. There had been a complexity there, he told me, that to this day no one understands:

SC: So, when it comes to behavioral measures, I mean, nothing would surprise me in terms of the way that the genetics has an influence on things.
NCN: What do you mean by that?
SC: I mean the complexity, the complexity and the way that a particular gene is expressed and interacts with other factors, other genes, environmental inputs . . . nothing would surprise me. So once again this notion of "this gene influences this behavior" is found to be an oversimplification in the first instance. Now we might find a few that have powerful effects, but I kind of feel like in psychiatry, if they were there, they would have already been found—you know, there's been so much effort put into this, in schizophrenia, bipolar disorder, and so on that if there were simple stories, I think that we'd be on to them. And almost all of those stories are still fairly muddled, at least that's my reading of it.

"Simple stories" were no longer on the agenda for Dr. Clark—he was adamant that there would be no discovery of a "gene for" alcohol addiction, or even a gene that strongly contributed to alcohol addiction. Researchers might be able to find a gene that made a small contribution to the overall behavior, accounting for maybe 5 percent of the difference in drinking behavior between individuals. And even in the case that a gene–behavior association could be stabilized, it would still leave many more unanswered mechanistic questions about how that particular gene produced the effect. What Dr. Clark aimed to produce, then, was an entry point or a shortcut into manipulating a larger phenomenon, not a firm fact. He explained to me:

> I teach [my students] that ultimately there's a complexity that we can only really scratch the surface on for most of the models that we use with animals. I think that the best thing that we can hope for is that we zero in on genes that do definable things in a mouse model, and hope that when that

> gene is examined closely in humans now we know enough about its function and expression. And maybe the alleles aren't the same in humans, but maybe the function of the gene is close enough so that the allelic diversity that exists in humans may tweak the systems in much the same way, and the model may have some transference. But certainly not that we will—I mean alcohol is a perfect example. There is no animal model of alcoholism that is comprehensive. You know, we have to model small pieces of it.

While Dr. Clark thought that they might be able to find specific genes that had an impact on well-defined subsets of the behavior (say, a gene that predicted whether a mouse would experience seizures when it went into alcohol withdrawal), others had different ideas about what would translate. Many researchers at Coast spoke about the possibility of generating new drug targets, either by identifying a specific molecule that had associations with drinking behaviors or by implicating a biological pathway that drug developers had not previously considered. Others thought that brain signatures might hold more promise for translation because they blended together genetic and environmental factors. The patterns of activity or receptor overexpression seen in the mouse brain might overlap with those of human drinkers, allowing clinicians to chart the course of the disease or identify people at risk. Another perspective was that researchers could contribute to the understanding of behavior by exploring how its boundaries and general properties changed under controlled conditions, such as whether an inherited propensity to drink could be modified by particular environmental circumstances. And yet other researchers, notably Dr. Smith himself, thought that they could best contribute by refining the existing tools of animal modeling so that future researchers could move forward with better instruments and do better science. As he put it, if he could provide more "bulletproof tests" for people to use, they then would be less likely to misinterpret behavioral data as they did their own investigations of gene function.

Coast researchers' characterization of the state of existing knowledge and future possibilities for knowing were to some extent echoed in the larger animal behavior genetics field. A news article in a 2008 special issue of *Science* on behavior genetics, for example, offers a description of the state of the field that aligns with Dr. Clark's view that researchers are currently only "scratching the surface" of the complexity of behavior:

> As scientists are discovering, nailing down the genes that underlie our unique personalities has proven exceedingly difficult. . . . All we really

> know so far is that behavioral genes are not solo players; it takes many to orchestrate each trait. Complicating matters further, any single gene may play a role in several seemingly disparate functions. For example, the same gene may influence propensities towards depression, overeating, and impulsive behavior, making it difficult to tease out underlying mechanisms. . . . Environment also plays a strong hand, bringing out, neutralizing or even negating a gene's influence. And genes interact with one another in unpredictable ways. (Holden 2008, 892)

As I will explore in more detail in chapter 2, even associations between a particular gene and a particular disorder have been challenging to produce, let alone definitive statements about what a particular gene or behavior *is*. As one researcher quoted in the *Science* article put it, behavior genetics research to date has produced "a trail of intriguing hints . . . but nothing that solidly replicates" (Holden 2008, 895).

All of these outcomes sound quite unlike what the scientific method is supposed to produce, at least if the lawlike products of the physics laboratory are taken as the rule. As Evelyn Fox Keller (2000) has argued, however, working from a vision of knowledge production that is based on the physical sciences devalues the type of knowledge work done in biology. Transposing the division between experiment and theory in the physical sciences onto biology creates the unflattering and misleading impression that biological research is either atheoretical or unable to purify its knowledge products into the desirable form of a universal theory. She argues that rather than trying to produce "models of" a phenomenon that make an enduring truth claim about how something works, biologists generate "models for" particular practical purposes. The goals and criteria for success in biology are different, and models are judged by how productive they are rather than how close they are to the truth.

Keller's argument about the form that theory takes in different fields could be extended to the form that facts and findings take in different research practices. Latour and Woolgar's (1979) famous exposition of the construction of a scientific fact, for example, directs attention to a particular kind of scientific fact—one that is "enduring, translocal, [and] transtemporal," as Park Doing (2008, 286) has put it. But establishing that TRF is a peptide with the sequence Pryo-Glu-His-Pro-NH2 is only one of many different kinds of knowledge products that might emerge from the laboratory. While animal behavior genetics depends on many facts of this kind for its functioning, this was not the kind of contribution that researchers at Coast expected they would be able to make. What they aimed for was knowledge that was partial, but useful. Even if researchers could not hope to produce an encompassing explanation of how addiction worked at a molecular level, they could at least provide some clues that might point clinical research in a more promising direction. And given what they saw as a present and pressing need for new treatments for psychiatric disorders, even partial and unstable facts were valued as useful clues in the hunt for treatment solutions.

CONCLUSION

In each of the instances I have described, researchers attended to complexity in the daily work of the laboratory as an epistemic problem, something that was salient because it created a barrier to knowing. The presumed complexity of both human behaviors and mice as higher organisms posed difficulties for using one to make knowledge about the other. Managing these problems of knowing required demanding experimental practices that involved attempting to tease out smaller, more stable behavioral units of analysis and controlling the laboratory environment all the way down to the smell of a researcher's shampoo. Coast researchers seemed to take it as a given that the ideal tools for studying behavior in its full complexity did not presently exist, and so for many, the tactics they used for creating entry points into the phenomena were a pragmatic compromise. Even for researchers who believed that behavior was in principle a predictable (if very complicated) phenomenon, the limits of their present knowledge of the relevant factors that produced each behavior and their ability to control for those factors meant that behavior often manifested itself in the laboratory as a stochastic phenomenon. For new researchers in particular, this was a set of epistemic expectations that could be disheartening or even paralyzing. Given the cultural value placed on enduring facts and theories, accepting that the best that they would be able to generate was a partial, unstable clue that might help other researchers in their work often involved a substantial readjustment of expectations.

The way in which researchers occasionally used "uncertainty" interchangeably with "complexity" at Coast further underscores the epistemological concerns that motivated complexity talk. Unlike "complexity," which appears at first to describe a quality of an object or situation, the term *uncertainty* points much more directly toward epistemological problems—situations where researchers believe their knowledge is incomplete or imperfect, and their ability to describe regularities or make predictions is limited. In using complexity to describe particular barriers to knowing behavior, researchers at Coast made complexity roughly synonymous with uncertainty.

To be sure, these epistemological stances were not disconnected from ontological stances on what kind of thing behavior was. If researchers assumed that behavior was an entity produced by the relatively straightforward action of a handful of genes, then many of the controls and precautions I have described here would have hardly been necessary. That behavior was an emergent, multifactorial phenomenon was a foundational assumption in the epistemic culture at Coast, one that animated all of the dynamics I have described here. Researchers disciplined the mouse's body, the laboratory environment, and their own bodies because they believed that behavior was complex. As Emily put it to me, "You can get easily overwhelmed by the complexity, but the bottom line is the complexity is the reality." But describing behavior as "complex" papered over a number of different assumptions and divisions about what kind of entity behavior was—notably, whether it was a nonlinear, interactive, system-level phenomenon or merely one involving many inputs; or whether the surprises and reversals that researchers experienced were an intrinsic property of behavior or a sign that they simply had not yet identified the major factors they needed to control for. Some researchers took interaction to be the rule; others thought that it would be important only in some cases. Many professed that it was likely that there were hundreds of genes involved in any given behavior, but some told me they secretly hoped that the number of truly important genes for their own behavior might be closer to a dozen.

Despite these divisions, invoking complexity at Coast did not provoke dissent and disagreement on these issues—quite the opposite. Instead, complexity was a unifying motif that organized many of the core epistemological beliefs and knowledge production practices of Coast researchers, such as their commitment to controlling the laboratory environment and their expectation that the knowledge they generated was partial. The flexibility of the term *complex* facilitated the erasure of ontological difference. Complexity talk was part of a shared epistemological discourse, one that left open the possibility that some "simple stories" might still exist, but that emphasized that the safer path forward was to assume that studies of the genetics of behavior would produce gaps, partialities, reversals, and uncertainty.

CHAPTER 2

Animal Behavior Genetics, the Past and the Future

One of my first interviews for this project was with Dr. Scott Clark, a collaborator of Dr. Smith's who also worked on the genetics of alcoholism. It was an interview that almost didn't happen. I had found Dr. Clark through an article he had coauthored with Dr. Smith, and Dr. Clark was within driving distance of my house so I e-mailed him for an interview and a tour of his lab. Dr. Clark did not respond to the first several e-mails, but I was determined not to let my project idea fail before it had even begun, and so I started calling his office. When I finally caught him on the phone, he was reluctant to talk to me but eventually agreed to an interview on one condition: that I would spend at least one hour with him. Any less, he said, and he would not be able to adequately explain the intricacies and caveats to the answers he gave to my questions.

When I finally met him in person, he took control of the interview almost as soon as I had turned on the voice recorder. He asked me to define "genetics" for him so that my own answer would be on tape. In the three-hour-long interview that followed, Dr. Clark systematically disabused me of what he correctly diagnosed as a view of genetics colored by my undergraduate training in molecular biology. He questioned me until he was satisfied that I had understood what it meant to study complex traits—quite a harrowing experience for an inexperienced ethnographer.

My encounter with Dr. Clark was not the only strange interview experience I had in the early days of this project. When I went to see Dr. Anthony Roy, a senior animal behavior geneticist with lively eyes and a quiet office tucked away at the edge of his Canadian university campus, he similarly took charge of our meeting. As I was about to ask him my first question, he abruptly pushed back from his desk and asked, "So you want to know how I got into this, the intelligence and heredity stuff?" He stood up and walked over to a bookshelf across the room, pulling out a worn copy of the *Harvard Educational Review*. It was his original copy of the 1969 issue containing Arthur Jensen's now-infamous article on heredity, race, and intelligence. In this hundred-page-long article titled "How Much Can We Boost IQ and Achievement?," Jensen (1969) laid out his case that intelligence was largely inherited and that the differences in IQ scores between racial groups were due to genetics. Jensen then concluded that educational interventions, such as policies designed to raise test scores in underperforming schools with predominantly African American populations, were not likely to succeed. Dr. Roy tapped his finger next to a highlighted passage on page 95 that he found particularly offensive. In this section, Jensen described the social ills that would result if welfare policies continued "unaided by eugenic foresight." While this article is undoubtedly an important one in the history of behavior genetics, I was confused as to why Dr. Roy thought that this was what I wanted to talk about. After all, I had e-mailed him asking to discuss his work on gene–environment interactions.

These were just a few of the moments in which the researchers I interacted with asserted control over their own narrative, attempting to show me how I should understand their research and its history. Researchers talked in very specific, prescribed ways about their animals and their experiments, and they also attended carefully to narratives about the larger field of animal behavior genetics. In my time at Coast, I learned quickly who—in their view—belonged to the field and who did not; what they considered to be the formative moments in the field's history; what lessons should be drawn from these historical moments; and finally, how practitioners should use these lessons to guide their path forward. The heterogeneous field of behavior genetics provided plenty of opportunities for didactically marking out difference—practitioners and viewpoints that fell outside of the Coast community often became foils that Coast researchers used to articulate their identity and commitments.

These historical narratives played an important role in forming a coherent epistemic community at Coast and shaping researchers' expectations about the future of their field. Stories about past experiences in the laboratory and key moments in the field's history served to organize individual practitioners' experiences into a shared story about engaging with complex behaviors, a story with common narrative themes about coming up against limits of their knowledge and readjusting their epistemic aims. These reconstructed pasts connected to an imagined future for the field, one where hopes for the creation of robust understandings about the genetics of behavior were pushed far into the distance. By rhetorically extending the time line on which meaningful results would be generated, researchers attached themselves to both hopeful narratives about the promise of genomics and pessimistic critiques of "genohype." And by connecting the field's history of heated public controversy to practices of cautious claims making, researchers articulated the need for an ethics of claiming, suggesting that those who failed to adopt their stances toward knowledge production were doing something both socially and scientifically dangerous.

THE HETEROGENEOUS FIELD OF ANIMAL BEHAVIOR GENETICS

Discussing the history of animal behavior genetics is a tricky prospect because the field is difficult to define. While some practitioners, such as Dr. Smith, readily identified with the "behavior genetics" label, others who were doing substantially similar experiments called themselves "neuroscientists" or "molecular biologists." Aaron Panofsky (2011, 2014) has described the behavior genetics field as an "archipelago" of loosely integrated communities. Animal behavior genetics, in Panofsky's description, is one "island" in a chain of distinctive research clusters that each has different affiliations with surrounding islands as well as with the "mainlands." Researchers at Coast saw themselves as part of the larger project of "behavior genetics" even though their experimental practices shared little overlap with psychological behavior geneticists who also identified with that label. Conversely, they shared common experimental questions and techniques with psychopharmacologists who would have not called themselves "behavior geneticists," even though they studied the genetic inheritance of behavior. They also had strong affiliations with the "mainland" of neuroscience, and many described their research to me as a subset of that larger field.

In his book *Misbehaving Science*, Panofsky (2014) argues that the unusual structure of the field is a product of its history of controversy. Panofsky points to the publication of Jensen's 1969 article as a turning point in the young field's history, one that shaped scientists' sense of their intellectual project and how they related to fellow practitioners. Faced with critics who boxed them in with Jensen and his purportedly racist research agenda, researchers adopted different strategies, with some defending Jensen and his right to intellectual freedom and others denouncing his research agenda as fundamentally flawed. These divergent strategies destroyed an implicit agreement that had existed up to that point: that race research was "scientifically intractable and socially destructive" (Panofsky 2014, 71). It also marked the beginnings of a split between the human researchers who wanted to defend the field, even if it meant defending Jensen, and the animal researchers who wanted to mark out Jensen's methods and claims as illegitimate.

Similar disputes flared up repeatedly over the next half century, creating an unusual situation where controversy was "persistent and ungovernable" (Panofsky 2014, 8) rather than a temporary state of affairs. The legacies of these incidents and practitioners' responses to them, Panofsky argues, are evident in the present-day structure of the field. Scientific work continues despite the presence of these unresolved controversies, but it does so in an altered form—the once-unified field has fragmented into different subgroups, practitioners have sought out intellectual homes outside the field, and they have even encouraged outside researchers to take up their questions and approaches rather than policing the field's boundaries.

Changes over time in experimental techniques have further contributed to the heterogeneity of the field. The 1990s were a particularly active moment of methodological change for mouse researchers, with two important techniques—genetic "knockouts" and "quantitative trait loci" (QTL) mapping—both introduced in the early years of this decade. The "knockout" technique enabled researchers to produce mice that had a nonfunctional copy of a gene by manipulating DNA in cell culture and then inserting the modified cells into a developing mouse embryo. After generating a mouse line that was missing a particular gene product, researchers could then compare these mice to unaltered "wild type" mice that had the gene intact. Although comparisons between mutants and wild-type animals had been used in behavioral genetic research programs for years, the knockout technique allowed researchers to target specific genes rather than generating mutations at random places in the genome or waiting for mutations to occur spontaneously. Better maps of the mouse genome and improved statistical techniques also provided researchers with new options for searching for genes associated with particular behaviors. QTL mapping combined newly available maps of genetic markers and statistical "interval mapping" techniques to allow researchers to identify regions of the mouse genome that varied between two populations. This method similarly held the promise of identifying specific genes that had a strong association with a particular trait, rather than simply providing a global estimate of the heritability of behavior.

These techniques, developed at the interface between classical and molecular genetics, academia and industry, drew researchers from different disciplinary backgrounds into unlikely relations with one another. As Sara Shostak

(2007) has argued, the mouse often functions as a "technology of translation" between diverse scientific communities by acting as a boundary object shared between social worlds and a focal point for standardization activities. The various ways in which researchers came to use knockout techniques to study behavior in the early 1990s illustrates this point. The first behavioral knockout study was published in 1992 by Susumu Tonegawa's laboratory (Silva et al. 1992), a researcher who was well known for his work on the genetics of antibody diversity but had no prior experience in behavior. Tonegawa got into the field, one of his colleagues recalls, because he was looking for a new scientific challenge after winning the Nobel Prize for his antibody work in 1987 (Kandel 2007). When a graduate student e-mailed him about the possibility of doing a project on his lab on memory, he took the student up on the offer.[1] Tonegawa's lab had ongoing projects using knockout techniques for studying the immune system (Mombaerts et al. 1991), which undoubtedly helped him set up a new research program using knockouts to study learning and memory.

Researchers with existing interests in behavior also found ways to incorporate new molecular techniques into their research programs. Eric Kandel, another Nobel Prize–winning researcher who had spent his career studying the molecular basis of memory storage using the sea slug *Aplysia*, published a paper describing the effects of a gene knockout on learning in the same year as Tonegawa (Grant et al. 1992). Kandel (2007) recalls that he was intrigued by knockout techniques because he thought they would make it feasible to study memory in higher organisms in the same molecular detail as he had done in lower organisms. To bring these new techniques into his laboratory, he hired a postdoc with expertise in mouse genetics from Cold Spring Harbor Laboratory and formed collaborations with two other research groups that had existing knockout mouse lines that were useful for his research agenda.[2]

The pleiotropy of the genes themselves also facilitated new collaborations. Researchers who created knockout mice found themselves drawn into new areas of research because of the unexpected results of creating such mutations, or fielding requests for collaborations from researchers in distantly related fields who were interested in the same gene. One of the first knockout experiments targeted a gene thought to be important in cancer, but the researchers discovered that knocking out this gene also resulted in severe brain defects (Thomas and Capecchi 1990). Some of the knockout mice Kandel used in his first experiments were similarly first developed to study suspected oncogenes (Schwartzberg et al. 1991; Soriano et al. 1991), but were useful for him because those same kinases were important in learning and memory. Researchers who had generated knockouts also took it upon themselves to find out if their transgenic mice might have interesting properties other than the ones they anticipated. Jacqueline Crawley recalls that in the early days of knockout experiments, she received many calls from researchers who had little familiarity with behavior but were interested in doing behavioral testing on their altered mice. By 1997, Crawley and her colleagues had tested over a dozen different knockouts or transgenics created by other research groups (Crawley et al. 1997), and Crawley (2000) recalls that she wrote the textbook *What's Wrong with My Mouse?* in part as a way to provide background information on behavioral testing to those she did not have time to collaborate with.[3]

Panofsky (2011, 2014) argues that many behavior geneticists advocated for deepening such intersections with other fields as a way of stabilizing the behavior genetics field. They encouraged researchers from other disciplines to incorporate the tools and techniques of behavior genetics into their own research agenda, creating a broader base of support for behavior genetics by "giving the field away" (Panofsky 2011, 304), as one of Panofsky's informants puts it. Affiliations with larger scientific fields such as molecular biology or psychology lent credibility and legitimacy to practitioners whose home field was frequently under siege. The outcome of these adaptive strategies, Panofsky argues, was an "inside-out" field, one where practitioners are primarily concerned with building credibility and scientific capital in neighboring fields rather than in their home discipline.

These relationships with neighboring fields looked rather different from the perspective of practitioners at Coast. Rather than happily "giving away" the tools of behavioral testing, they felt as though those tools were being taken away from them and used inappropriately. When I first started fieldwork at Coast, I inadvertently provoked a flurry of well-worn complaints about non-behaviorists and behavioral testing by telling the grad students that I was interested in the ways that researchers with different disciplinary backgrounds approached experimental design. One grad student explained what the situation looked like from their point of view: in order to get a paper published on a new mouse mutant in high-profile journals such as *Nature* or *Science*, researchers needed to pull together multiple kinds of evidence, and behavioral testing was often the component where those publishing the high-profile papers had the least expertise. Researchers who lacked behavioral training did not know how to properly run these experiments, she told me. Even if they did manage to do the experiments well, they were "not very thoughtful" about how they interpreted the data because they did not understand the origins of these tests or their limitations.

When I asked Dr. Roy in my interview with him about how molecular biologists might approach behavioral test development, he also reacted strongly: "Well, I wouldn't let the molecular biologists have anything to say about it! [laughter] Because what do they know about [behavioral tests]? I mean, this is something where the behavioral psychologists have to get together and provide good, validated tools that the molecular people can then use." In his opinion, those outside the field would only be able to profitably make use of behavioral tests once those tools were stable, which was presently not the case.

Researchers found it difficult to exert control over the tools of the field, however, because they saw the barriers to entry to behavioral testing as low. Unlike molecular biology, where the equipment needed was expensive and it was quite difficult to get new techniques to work, behavioral tests were inexpensive and (seemingly) easy to run. A researcher interested in testing mice for anxiety could buy an "elevated plus" maze for relatively little money, and because the test involves simply measuring how much time a mouse spends in various areas of the maze there was no way not to produce data. This created an asymmetry where researchers at Coast felt that outsiders were free to take up behavioral techniques themselves, whereas they themselves needed to seek out experienced collaborators if they wanted to take up molecular techniques. "I can go down the hall and borrow a plus maze for a week," one grad student told me, "but I can't go and borrow a sequencer."

The difference between Coast researchers' views on the heterogeneous nature of the field and those of the behavior geneticists that Panofsky (2014) describes may be attributable to animal behaviorists' relative disenfranchisement in both behavior genetics and molecular biology. Panofsky observes that while animal researchers played a central role in founding the field in the 1950s and 1960s, they became increasingly marginalized as the field developed.[4] By the time I arrived at Coast in the late 2000s, behavior genetics no longer seemed to be a fertile intellectual home for the animal researchers I interacted with. Dr. Smith told me that he remained "committed to the idea of behavior genetics" and still maintained his membership in the Behavior Genetics Association (BGA), but he had not been to one of their annual meetings in more than a decade. When I attended the annual meeting of the BGA in 2008, there was only one oral presentation by an animal researcher on the agenda, who lamented their lack of representation when he took the podium to speak.

Animal behaviorists' relations with molecular biology were also fraught. To veteran behavioral researchers, knockout studies seemed to be yet another in a long line of colonizing moves made by molecular biologists. Pnina

Abir-Am (1992, 165) argues that molecular biology was perceived as a threat by organismal biologists and biochemists in the 1960s because it was "redefining, and hence appropriating, many concepts, both central and peripheral, around which the 'classical' disciplinary monopolies were constituted." The same could be said of developments in the 1990s, when molecular biologists appeared to be redefining what it meant to do "the genetics of behavior" in terms that worked to their advantage. It was not the selective breeding experiments that animal researchers had been doing since the 1950s that benefitted from increased enthusiasm and funding for genetic studies, but projects that employed molecular techniques. Steven Hyman (2006), director of the National Institute of Mental Health from 1996 to 2001, recalls that funding for genetic studies in animal models "markedly increased" during his tenure, but that most of this investment went into knockout mouse models and large-scale mouse mutagenesis projects.

The uneven terms on which veteran animal behaviorists felt they were competing thus intensified rather than weakened Coast researchers' desire to draw boundaries and maintain control over their experimental techniques in the heterogeneous field. While they felt they lacked the ability to physically maintain control over inexpensive behavioral equipment or the creation of behavioral mutants, one thing that they could exert control over was the interpretation of those experiments, and over broader narratives about where the field had come from and where it was going.

CAUTIONARY TALES FOR KNOWLEDGE PRODUCERS

Narratives about past experiences in the laboratory, as we have already seen in chapter 1, were a resource that researchers at Coast used to shape collective research practices and expectations about research products. Dr. Smith's story about Gerald McClearn's failed selective breeding project and the graduate students' reflections on their undergraduate research projects are examples of how researchers used narratives about past experiences to justify particular methodological stances in the present day. These narratives should not be understood as mere rhetoric but as techniques through which researchers materially shaped the field's practices. As Nik Brown and Mike Michael (2003) have argued, researchers' descriptions of past events and future trajectories are a means of enrolling people and resources into the futures they describe. These shared visions of a technology or a field's time line can serve to marshal resources, coordinate activities, or manage uncertainties. Researchers who believe that their work is still a long way from application, for example, might be less likely to seek connections with clinical researchers or drug companies, thereby creating the conditions to realize the future they foresee. In this way, Brown and Michael argue that actors' histories create the sociotechnical networks that support the particular futures they envision.[5]

Many of the most frequently recounted narratives at Coast contained methodological lessons, and in particular lessons about what not to do. One "cautionary tale" I heard repeatedly during my time there—in the introductory behavior genetics class, from current and former members of the Smith Lab, and from Dr. Smith himself—was the story of the disappearing drinking difference. This story recalled a series of knockout experiments conducted in the Smith Lab more than a decade earlier. The transgenic mouse they were working with had been created by another laboratory, and Dr. Smith had obtained the knockout so that his lab could test its drinking behavior. Here is Dr. Smith's version:

> We were testing a knockout for drinking alcohol, and they drank a lot of alcohol compared to wild types. And we did it three times, and we got the same answer, three times. And we said, "Wow, that's pretty good." So we published it, and in the meantime we'd gotten more mice to replace ours because they'd gotten too old to breed anymore. When we redid the drinking study, and there was no difference [between the knockouts and the wild types]! So we tested them again, and we got no difference again! And we were scratching our heads trying to figure out what went wrong, and we screwed around with the parameters for a year trying to figure out what was different, and we'd get [the difference] sometimes, we wouldn't get it other times . . . one study we'd get it in males, the next one we'd get it in females . . . it was just really weird. And then those mice got too old and [our collaborator] sent us a third batch of mice, and lo and behold, boom! There it was again, just really big.

Despite strong results in their initial experiments, the Smith Lab later encountered problems replicating their own results. Several other research groups later studied the same knockout, and many could not find the difference in drinking that Dr. Smith initially reported. Dr. Smith recalled that this situation was initially baffling, and there seemed to be no reasonable explanation for why the drinking difference disappeared, until the effect reappeared with the third batch of mice.

At that point, Dr. Smith suspected that something about the breeding practices of the research group supplying them with the knockout mice might account for the difference. A phone call to the other lab revealed that they were using different substrains of mice for creating their knockouts. While all of the mice Dr. Smith's lab had received were missing the same brain receptor gene, other genes in the mouse's genome changed depending on which mouse substrain the other laboratory was using for breeding at the time. Dr. Smith went on to demonstrate through what he described as a "long and tedious breeding scheme" that if the knockouts were made using one substrain, then they would drink more, but if they were made using another substrain, the knockouts would drink the same amount of alcohol as the wild-type mice.

The story of the disappearing drinking difference conveyed a methodological point about what became known in the field as the genetic "background" effect (Gerlai 1996). The initial techniques used to create knockouts involved blending the genomes of several different mouse strains together. Researchers were at odds about how much this variation in the genetic background of the mouse mattered for interpreting the results of knockout experiments. Dr. Smith recalled that it was immediately obvious to him that such genetic variation had the capacity to alter a mouse's behavioral profile, but other researchers were not convinced that such small variations could overcome the seemingly large effect of removing an entire gene product.

Retelling this story about the disappearing drinking difference was a way of drawing attention to the potential problem and encouraging fellow researchers to change their experimental practices. And it was an effective one. Dr. Charles Westin, a former student of the Smith Lab who also recounted this story to me, said that these events made him and everyone else in his lab "acutely aware" of the importance of controlling for the genetic background, even before examples of background effects were published. This story was also an effective way of communicating the importance of creating controlled conditions for studying behavior more generally because it showed how a seemingly small variation—the handful of genes that differ between mouse substrains—could lead to the dissolution of a valuable experimental result. At Coast, retelling this story lent a sense of urgency to the task of implementing controls and taught researchers to regard results from what they saw as poorly controlled experiments with suspicion.

Cautionary tales also served to reinforce epistemic expectations, emphasizing the limits of researchers' capacities to control the laboratory environment and to know behavior. In Dr. Smith's retelling of the disappearing drinking difference story, the moral was less about the background effect and more about the limits of experimental control. "That was one," he concluded, "where, you know, you try to be as careful as you can and you try to think of as much stuff as you can, but you can't think of everything!"

Other cautionary tales that circulated at Coast were quite vague in their prescriptions for experimental practice but powerful in their messages about the epistemic limitations of animal behavior genetics. One such story concerned a problem the Martin Laboratory had encountered, involving the disappearance of the "stimulation effect." Ava, an outgoing student from Texas who had just joined the Martin Lab as these events were unfolding, recounted the story to me in an interview: "We had this situation last year, two years ago where we couldn't get a stimulation effect when our mice were given methamphetamine. Many strains of mice will show heightened locomotor behavior when they're given a dose of meth compared to the saline animals. Well, all of a sudden, like we weren't seeing that in our animals. We'd seen it multiple, multiple times. They didn't show this acute response, and it was just kind of like, what the hell is going on?"

Unlike the story of Dr. Smith's knockout study, the result that disappeared here was not even a new experimental finding but a well-known baseline result that was part of the initial process for setting up more complicated experiments. The problem persisted for several months, and the Martin Laboratory investigated many different possibilities that might explain the strange results. They tried having different technicians run the experiment. They used different bottles of methamphetamine. On the suggestion of the controlled substances authority at Coast, they even tested to see if the stock bottles of methamphetamine actually contained the drug, in case someone had been stealing it from the laboratory and replacing the liquid in the bottles. The stimulation effect eventually returned, but the source of the problem was never definitively established.

Without a clear resolution to the story, researchers drew different conclusions about methodological lessons to be learned as they retold it to me. Ava suspected that the problem was that the bottle of drug that they were using was somehow ineffective, although Dr. Martin noted that the results of the laboratory analysis had shown that the methamphetamine was pure. Dr. Martin thought that the changes might be due to differences in the early environments of the mice purchased from a commercial supplier versus mice bred in their own facility. But regardless of the specific experimental factor each individual focused on, all of the researchers who retold the story to me emphasized both the importance of experimental control and the limits of their present knowledge. Chloe, a student who rotated through the Martin

Lab during her first year while this problem was taking place, concluded her retelling of the story as follows:

CHLOE: It just shows you sometimes the science isn't that—sometimes things don't work out as planned.
NCN: Sometimes it's not as controlled as you would like it to be after all, eh?
CHLOE: Exactly.

CULTIVATING CAUTIOUS CLAIMING

Along with emphasizing specific methodological problems and solutions, then, cautionary tales served to emphasize the unfinished status of the behavior genetic knowledge produced thus far and the need for researchers to be careful about the scientific claims they made. These stories amplified and extended the messages researchers at Coast received about complex phenomena and their associated epistemic problems. The young researcher who heard these stories about past experimental problems learned how to conduct an experiment that met her local community's understandings of behavior, and absorbed expectations about what types of claims about behavior were acceptable.

Another one of the ways in which Coast researchers' stories imparted these messages was by setting up a contrast between the cautious claimers who turned out to be on the right side of history, and those whose positions looked obviously wrong in retrospect. These stories illustrated the types of scientific claims that were likely to stand the test of time—or not. Stories about the "molecular revolution" in the 1990s were often told in this way. These new techniques created openings for claims about genes and behavior with an unprecedented degree of specificity, and stories about these techniques made excellent fodder for cautionary tales both because their claims looked audacious in retrospect and because of who had made those claims: the researchers who created the first knockout mice were largely outsiders to behavioral research like Tonegawa, and the "gene for" claims they had made were the antithesis of the present-day emphasis on complexity at Coast.

Prior to the methodological changes that took place in the 1990s, the kinds of claims that researchers could make were fairly general: they could make statistical estimations of how many genes might be involved in a particular trait, but they did not know which genes those might be or have any methods available to identify them. After the "molecular revolution" (as they called it), researchers could implicate specific genes in specific behaviors. Tonegawa and his coauthors were quick to note this in the publication describing their first behavioral knockout study. They wrote that previous research studies comparing different inbred strains of mice had generated correlations between levels of particular kinases in the brain and performance on learning and memory tasks but that the differences identified through those experiments were "clearly not the result of differences in a single gene." They argued that their kinase knockout experiment, in contrast, demonstrated the "selective but drastic" impact that a single genetic change could have (Silva et al. 1992, 210).

Coast researchers' stories about these events and these practitioners resembled a narrative "status degradation ceremony" (Garfinkel 1956), where molecular biologists started out as confident, powerful entrants onto the behavioral research scene and ended up humbled by the complexity of behavior (and by the knowledge of veteran animal behavior geneticists). Dr. Clark's recounting of these events, for example began as follows: "What we started seeing in the early years of the genetic engineering revolution with mice was study after study where they would look at aggression using a simple task. And what we started to see was that pretty much every gene that was knocked out could be called an aggression gene. But then there were some notable failures to replicate, where the same knockout, when sent to somebody else's lab, they would find something different."

Dr. Smith's difficulty in reproducing his knockout study was only one of several such failures to replicate. In another case, two independent research groups (one of which was Tonegawa's) created knockouts of the dopamine D1 receptor. While one group found that the mutation produced an increase in the locomotor activity of the mice, the other group found no such behavioral difference (Drago et al. 1994; Xu et al. 1994). At Columbia University, a research group reported two different behavioral profiles for their serotonin receptor knockout mice: In early studies, they found that mice lacking the receptor showed no difference in the open field test (a test of anxiety), but in later work, they found that these same knockouts showed more anxiety behavior in this same test (Ramboz et al. 1995; Zhuang et al. 1999). As in Dr. Smith's knockout story, Dr. Clark recalled that it was immediately obvious to him that changes in the genetic background of knockouts could account for these differences. He said:

> So as psychologists, we're looking at this, and we know something about genetics, and there are really two hypotheses about a failure to replicate when somebody's taking a knockout that was created in this one lab and now two different labs are testing it and they don't find the same thing. One

> is this genetic background thing, and on that count, in the early years, and to some extent still, molecular biologists have been horrible geneticists. . . . The background effect hypothesis was something that they didn't want to hear, they got beat over the head with it, there were negative reactions to the psychologists pointing this out to the molecular biologists.

When recalling the debates around knockout experiments, researchers at Coast emphasized the differences between themselves and the assumptions they made based on their background in psychology and animal behavior, and practitioners trained in molecular biology. They portrayed molecular biologists as overly confident in their initial interpretations of knockout experiments, only belatedly coming to the realization that they were claiming too much on the basis of their experimental data. Once the aforementioned issues with interpreting and replicating the results of knockout experiments became too evident to ignore, behaviorists contended that molecular biologists were then overly pessimistic in their interpretations of these problems. The molecular biologists of animal behavior geneticists' stories always claimed too much or too little, but rarely did their claims end up being just right.

Some stories blamed molecular biologists' seeming inability to properly moderate their claims on the "gene jocks'" tendencies toward self-aggrandizement. A more charitable narrative was that they simply held different assumptions about how genes worked and how to study their action. In explaining why the molecular biologists ended up on the wrong side of history, the practitioners retelling these stories reinforced messages about the complexity of behavior and the epistemic stance required of those who studied it. Dr. Clark, for example, emphasized differences in the way that molecular biologists and veteran behaviorists understood gene action:

> The perspective from the point of view of molecular biology is sort of "one gene at a time," and you're not faced at the outset with the issue of control of particular phenotypes by multiple genes, multiple sets of genes. And so that aligns itself with the medical model that historically had tried to find a gene for a disorder, right? That contrasted very much with the field of behavior genetics as it developed, which was immediately acutely tuned in to this complex control system likelihood. Anything that we were likely to study, if it had a genetic component, would likely be controlled by multiple genes, perhaps interacting, perhaps not, who knows?

Dr. Steve Fortin, a Canadian animal behavior geneticist, made a similar argument. In his narrative about knockout debates, he pointed out that the molecular biologists' initial position—that the background effect was a non-issue—seems more reasonable if one takes into account that they were operating under the assumption that the effect of knocking out a gene would be pronounced. He explained:

> For [molecular biologists], a lot of these questions were less problematic. For example, if you knock out a gene which influences, let's say, limb development and you end up with a mouse that doesn't have any legs, OK? That's such a major developmental alternation that these background genes and modifying effects or compensatory changes or the so-called flanking allele problem are basically irrelevant. It's such a robust mutation that it doesn't really matter. But for us, behavioral neuroscientists and behavioral geneticists, we cannot think like that. . . . We have to be a bit more worried about these seemingly minuscule, negligible genetic effects. For us, these effects are real and they're not negligible.

As in Dr. Clark's narrative, Dr. Fortin concluded that molecular biologists were wrong about the background effect because they did not have an adequate appreciation of the complexity of behavior. Therefore, they made overly simplistic claims that in time were shown to be unsupportable. In these narratives, those who were able to adjust their expectations accordingly were then presumably able to return—humbled—to participation in the animal behavior genetics project.

Occasionally the antagonists in Coast researchers' historical narratives were not researchers from other disciplines but their own past selves. These stories modeled how an ideal researcher should deal with such readjustments of understandings and expectations. At one scientific meeting that I attended with Dr. Smith, he was receiving a lifetime achievement award, and the presentation ceremony took the form of a roast. One of Dr. Smith's longtime collaborators presented a lighthearted narrative of his research trajectory, taking the opportunity to poke fun at the "wrong turns" in Dr. Smith's "scientific quest." The commentator described one project that Dr. Smith had embarked on as a postdoc: Dr. Smith was looking for what is known as a bimodal distribution in his breeding experiments, where there are two distinct subgroups within a larger population (such as a cluster of heavy drinkers and a cluster of light drinkers). The idea, the commentator said, was a sound one. Evolutionary theory and population biology predicted that when a bimodal distribution appeared, it suggested that the two subpopulations had diverged genetically, and so bimodal distributions could be used as an indicator to detect genetic difference. The problem was that for such a plan to work, a gene would have to account for 82 percent of the variance in a trait for a bimodal distribution to appear. This comment drew a mixture of chuckles and gasps from the audience, because by contemporary standards this would be a project obviously destined for failure—a gene accounting for 5 percent of the difference in a population is now considered quite large, and researchers expect that most gene effect sizes will be in the neighborhood of 1–2 percent (Flint et al. 2005).

The way in which the commentator elaborated the rest of Dr. Smith's scientific history makes clear the didactic function of recounting this past for present-day researchers. Dr. Smith's commentator teased him for his foolish, youthful assumption that a single gene could have such a pronounced effect on drinking behavior, but he went on to describe Dr. Smith as a "consummate scientist" because of his willingness to "destroy his own data." He recounted several instances in Dr. Smith's career where he had complicated existing scientific stories with his research, sometimes even his own stories, as in the case of the disappearing drinking effect. Dr. Smith's paper showing the association between lowered drinking behavior and the knocked-out brain receptor had been published in a high-profile journal and received a good deal of attention. But not content to let such a simple scientific story stand, the commentator said, Dr. Clark showed in a series of other papers (notably, published in much less high-profile venues) that this seemingly straightforward relationship between missing gene and altered behavior depended on the genetic background and the way that the behavior was measured. "And now," the commentator concluded with a smile and a feigned sigh, "we have just another complex trait, like so many others in the literature." Dr. Smith echoed this narrative arc when commenting to me about the event later on. "While I made a career I have really enjoyed out of this overall project," he said, "I can't believe how naïve I was to think one could ever accomplish that."

AN ETHICS OF CLAIMS MAKING

Similar narrative elements were to be found in researchers' stories about another set of historical events: debates around race, intelligence, and heredity. Many of the senior animal behavior geneticists whom I interacted with began their training in the field in the late 1960s and early 1970s, at a time when debates about this topic were erupting into highly public and politicized controversies in the United States. Dr. Roy, who had just finished his graduate training in the late 1960s, vividly recounted to me the moment when Jensen's article was published:

> I did a postdoc at the Institute for Behavioral Genetics at Boulder, Colorado, where they work with mice. So I began to learn about inbred strains and behavioral testing in mice because I had just done dogs before, so it was a lot to learn, a new species. Well, I'd only been there a couple months when—and like I say, I was really more of a pure scientist, interested in the biological basis of learning and memory—and all of a sudden, boom! This bomb was dropped into the middle of our work called the *Harvard Educational Review*, an article written by Arthur Jensen on race, genetics, and intelligence.

In the wake of Jensen's publication, fistfights broke out at academic meetings, protestors gathered on several university campuses, high-profile scientists denounced the methods and findings of the field, and some behavior geneticists even received threats against themselves and their families.

For many practitioners, it seemed that the wounds from these debates had barely healed when they were hit by another wave of public controversies in the 1990s. The 1994 publication of *The Bell Curve*, an eight-hundred-page tome by Richard J. Herrnstein and Charles Murray, reopened race, heredity, and intelligence debates. Like Jensen before them, Herrnstein and Murray (1994) drew together a variety of data sources to argue that IQ testing is a useful measure of general intelligence, that there are racial differences in intelligence, and that American society was gradually diverging into a "cognitive elite" and a cognitive underclass with higher rates of crime, poverty, and unemployment. The book received national news coverage, selling half a million copies, and it generated a flood of criticism both from scientists and public commentators.[6]

Less than a year after the publication of *The Bell Curve*, a speech at the BGA annual meeting from the society's outgoing president, Glayde Whitney, brought the race and heredity debates even closer to the core of the field. At the closing banquet celebrating the twenty-fifth anniversary of the society, Whitney argued that it was time for the field to move toward a new agenda of studying differences between human racial groups. He presented data on the murder rates in countries and cities with different "racial compositions" in their populations, and argued that it was a reasonable hypothesis to investigate whether these differences in the murder rates could be attributed to genetic differences between races. Some members of the audience, including a few seated at the executive table, walked out of the banquet in protest.

Even for younger researchers who had no firsthand experience of these events, these controversies were still very much a part of their own sense of the field's history. When I was out for lunch one day with Emily, I mentioned to her that I was having trouble working out my travel schedule after my stay at Coast ended. I wanted to go to the annual meetings of both the BGA and the Research Society on Alcoholism, but they overlapped. Emily responded that this was often the case, but it did not really matter, because no one from Coast really went to the BGA meeting anymore. She explained to me that there was a split in the association years ago after the president gave a "Bell Curve–like" speech at one of the meetings that had people "literally yelling at each other." She told me she had heard her account of these events from Dr. Smith, who had stopped attending the BGA meetings after Whitney's speech.

Like stories of the molecular revolution of the 1990s, narratives about these events highlighted particular methodological problems and transmitted more general messages about the need for researchers to exercise caution when formulating scientific claims. The antagonists in these stories were once again scientists who were reckless in their interpretations of experimental data. But this time, rather than simply being proven wrong over the course of time, these scientists' audacious claims brought protesters to the field's doorstep. These incautious claimers were also closer to the core of the field. The Whitney incident seemed to make an especially strong impression on researchers at Coast because unlike the molecular biologists, he could not be easily othered: Whitney was a veteran behavior geneticist working on the genetics of taste sensitivity in mice, well respected for his work before he "went off the deep end," as one researcher put it to me. Stories about the race, intelligence, and heredity controversies heightened researchers' sense of the seriousness of their responsibility to make careful claims, turning claims making into an activity with moral significance. Language emphasizing the ethical importance of formulating scientific claims about genes and behavior appeared not just in these particular narratives but also throughout everyday conversations at Coast.

Because many practitioners responded to Jensen, Herrenstein and Murray, and Whitney by writing detailed critiques of their methods, stories about these controversies offered ample opportunity for making specific methodological points. Dr. Smith used one of these texts to teach the concept of heritability to the introductory behavior genetics class. The text was a discussion of heritability analyses written in 1994 by Douglas Wahlsten, an animal behavior geneticist who was well known for being an outspoken critic of arguments about heredity and intelligence. The article explained the concept of heritability through a detailed analysis of its limitations, such as the problem of taking interaction into account in a model designed to separate out sources of variation, and the problem of using heritability scores generated in one context to understand another. "In view of all this," Wahlsten (1994a, 252) concluded at the end of the article, "I would feel more secure riding a three legged moose over thin ice than relying on a heritability coefficient to help me understand the origins of individual differences or predict future levels of intelligence." The article was followed by a defense of heritability analyses by another behavior geneticist, and in his reply to that defense, Wahlsten elaborated on his reasons for being critical of the technique. He described his own early career experiences as a student of the leading figures in heritability analysis at the time, and how he gradually came to the opinion that the technique was not only theoretically problematic but was "actively misleading" to those outside the field (Wahlsten 1994b, 265). When Jensen's article was published, Wahlsten recalled that he saw clearly how problematic the application of these theories to human society could be. "What seemed innocent enough when used [for breeding programs] on the farm," he wrote, "took on sinister overtones in political debate" (Wahlsten 1994b, 265).

The stories other practitioners told me about their experiences of these events echoed Wahlsten's narrative about developing a sense of personal responsibility for managing the potential danger of claims about genes and behavior. Some of Dr. Smith's colleagues also used the race, heredity, and intelligence controversies as part of their scientific pedagogy. A former member of the Smith Lab, who was just getting his start in the field during the controversies of the 1990s, told me that he assigns the text of Whitney's speech in all of his classes as a reminder of the potential for "racist misuses" of behavior genetics research. For Dr. Roy, the Jensen controversy disrupted his identity as a basic scientist whose work had little to do with contemporary social problems. From that point forward, he told me, he devoted a substantial proportion of his career to penning methodological critiques designed to rein in what he saw as inappropriate and dangerous claims. Dr. Clark, in contrast, emerged from these same controversies with an even greater conviction that he was a basic scientist whose work should not be seen as having immediate implications for humans. These events demonstrated to him that it was difficult, if not impossible, to craft responsible messages about behavior genetics work for public consumption. So, he responded by attempting to keep himself and his work out of the media as much as possible. He told me that throughout his career, he had avoided the "attention seeking" press releases and media coverage that he believes some other behavior genetics labs generate, because of the risk of misinterpretation: "All those things that have social implications that can immediately start—you can immediately start relating them to racial and group differences in humans, and that's dangerous ground. Even if you come to the material in a completely objective way, there are going to be attempts to distort what you have to say, and draw conclusions beyond what the data would permit." Although Dr. Roy and Dr. Clark adopted opposing strategies for public communication, their stories describe a common experience of developing an increased sensitivity to the dynamics of scientific claims making. In retelling these stories, they amplified warnings to their fellow scientists about the social danger of scientific overclaiming.

Researchers at Coast even occasionally told stories about their field that linked it back to histories of eugenics. Panofsky (2014) has argued that the practitioners who coined the term *behavior genetics* to describe their work in the post–World War II period did so deliberately to try to distance their work from eugenics—although as Diane Paul (1998) has shown, their research was supported by the explicitly eugenic agendas of leaders at the Rockefeller Foundation. Unsurprisingly, most actors' histories of behavior genetics downplay the connections between early twentieth century eugenic thought and postwar research, but a few did insist on making these links. Douglas Wahlsten, for example, published a critique in the journal *Genes, Brain, and Behavior* of the Nuffield Council on Bioethics' (2002) report on ethics and behavior genetics. He accused the authors of the report of "airbrushing" the concept of heritability by not making clear that the heritability analyses practitioners use today were developed in the context of eugenic agendas and formed part of the justification for eugenic sterilization programs that continued up until the 1970s in the United States and Canada (Wahlsten 2003).

These actors' histories cultivated an ethics of claims making at Coast, one that imbued to particular ways of speaking, behaving, experimenting, and predicting with moral significance. Drawing histories of scientific racism into stories about sloppy methods and hyperbolic scientific claims deepened the sense of ethical responsibility researchers felt with respect to their knowledge production practices. I was often surprised by the kinds of methodological problems that researchers at Coast described as morally significant activities. One morning Liam, a Coast graduate student who usually was quite calm and collected, came in upset because he had discovered that the light timer was broken in one of his mouse housing rooms. He had stopped by the room to check in on his mice before leaving in the evening and noticed that the light was still on when it should not have been, and he didn't know how to manually reset the lights because the building staff controlled them. He was set to run an addiction experiment that morning with those mice, which he had spent two weeks preparing for, but he told me he knew already that he "really couldn't say anything" about the test he was running because he knew their circadian rhythms were "messed up." I asked him whether he might still get good data in spite of the lighting malfunction, and he said he might, but it would not matter. "To be an honest scientist," he told me dejectedly, "I really can't use this data anymore." The adjective "honest" stuck with me. I was surprised that Liam would discard a data set that he had spent weeks preparing to collect because of a change in the housing room that might not even register in his results, and that he would imply that it was dishonest not to do so. Breakdowns and unexpected events happen all the time in laboratory work, after all. But under assumptions of complexity, testing a group of mice exposed to different environmental conditions amounted to bringing into being a different version of addiction behavior. And in a culture of cautious claiming, putting a potentially misleading statement about the genetics of addiction out into the world was irresponsible, and potentially dangerous. The ethical thing to do, then, was to start again from the beginning.

ALTERING THE GENOMIC FUTURE: THE LONG HORIZON OF KNOWLEDGE PRODUCTION

In their work on the sociology of expectations, Brown and Michael (2003) have argued that one function of recounting scientific pasts is to draw practitioners together in a shared vision of where their field is heading, creating a road map for the future by extrapolating from the contours of the past. The role of these future-oriented narratives in the contemporary life sciences has attracted much attention in recent STS work. Michael Fortun (2008, 43) has argued that genomics is inescapably promissory—if once biotech start-up companies survived by making products, today they "survive by making promises." The genomic future that companies, scientists, and politicians envision is volatile and uncertain, but it is a hopeful one—one where it will be easier, cheaper, and faster to identify genes and use them to manage future risk of disease. The increasing volume of genetic information and especially the increasing speed at which researchers can obtain and analyze this information are central to this vision. Fortun (2008, 44) writes of the deCODE genetics project that "if the vast majority of [their candidate gene] leads never went anywhere, that surprised no one. You simply returned to the fire hose and drank again."

Quantity and speed acted as counterbalances to the uncertainties of genomic discovery, securing the promise of eventual payoffs even in the face of a present filled with failures. Analysts have taken pains to avoid dismissals of this persistent optimism as mere hype, an exaggeration of the actual potential of genome science, instead showing how these hopeful speculations play a role in creating the biotechnological futures they predict (Brown and Michael 2003; Fortun 2001, 2008; Hedgecoe 2010; Hedgecoe and Martin 2003).

Coast researchers' vision of the trajectory of animal behavior genetics in many ways resembled this narrative, but there were important differences. Their work was deeply embedded in a hopeful vision of what genomic information might become. Although precisely what would eventually travel out of the animal behavior genetics laboratory and what clinical form it would take was loosely and multiply articulated, the promise that their work would eventually transform human health underwrote all of the research that took place at Coast. Practitioners' stories about the path to producing this clinically relevant knowledge also followed a familiar promissory pattern: someone makes "inflated" claims about the promise of knockout experiments or breeding programs to generate bimodal distributions, those experiments fail to deliver and hopes are dashed, and scientists then rebuild a more moderate, stable version of their research program. This narrative arc is pervasive in stories about technology development—so much so that a technology consultancy firm has branded their own graphical representation of this so-called "hype cycle" and claim to be experts in "hype cycle research methodologies" (Gartner Inc. 2016).

But in contrast to the emphasis placed on speed and volume in mainstream genomics narratives, some of the notable features of the future that researchers at Coast envisioned were its long time horizon and slow pace of accumulation. Despite what senior behavior geneticists described as "revolutionary" advances in technique that they had experienced in their careers, they often described the field as one that was "still in its infancy." The stories they told about what might come next sometimes measured in years but often extended decades into the future, even well past researchers' scientific lifetimes. In the absence of expectations of a rapid pace of knowledge accumulation that would smooth out the uncertainties of genomic research, Coast researchers' narratives were optimistic about the long-term future but pessimistic about the near term, as though they expected to spend their whole lives in the "trough of disillusionment" of the branded hype cycle.

Graduate student Emily's vision for the future of the field shows how extending the time line of clinical translation markedly changed the tone of familiar narratives about genomics. The type of future scenario she described—one where her research had been translated into tests that could be used to counsel parents about their child's risk factors for behavioral disorders—is a clinical story that is pervasive in contemporary genomics. But by pushing this scenario as it applied to behavioral disorders a hundred years into the future, it transformed from something that seemed plausible into something that seemed like science fiction. She said:

> Wouldn't it be cool if in a hundred years, everyone sat down with a genetic counselor, they did some testing, and they said, "Your baby is predisposed to X, Y, and Z; here are the environmental things that you could do to protect them against expression of this disease." So you'd get a better shot. I'm not an advocate of changing the genome, I'm not an advocate of genetic modification in humans, but you could try to protect people. You could protect people, because they have shown that even if you have genetic predispositions, by being raised in certain environments you will either have less of the disease or it won't show up at all.

Emily's imagined future is exactly the type of scenario that investors, grant funders, and would-be patients are regularly asked to imagine on a five- to ten-year time line, or even as something that could exist today. Indeed, direct-to-consumer genomics companies are already using genetic testing to give consumers information about their predispositions for particular diseases and advise them on how to modify their diet and behavior accordingly. But Emily was steadfast in her time line for behavioral research: "Even if I were doing amazing, wonderful work in alcoholism," she told me, "it would still be like fifty years until we saw anything."

It was jarring to consider together the clinical orientation of the animal modeling work conducted at Coast and practitioners' assertions that they were basic researchers whose work was still far from application. Although researchers frequently told me that the aim of their work with animals was to produce findings with relevance for human health, they were equally insistent that actual treatments were "a long way off." When I interviewed Alex, a senior graduate student in the neuroscience program, I was surprised to find that he said very little about the clinical relevance of his work, especially because the word *translational* was sprinkled liberally throughout the description of his laboratory's research on the department website. When I asked him directly about the applicability of his research to human health, he was reluctant to talk about it:

ALEX: How close to addiction is [the animal model I'm using]? I mean, it's not close at all. Yeah, certainly there's some brain changes going on, but do they match the changes that would have gone on if I'd given them a thousand injections, or that the humans who do cocaine every day for seventeen years? No, it doesn't. . . . My stuff is more—honestly I'm more interested in the memory aspect of it in my research. This association that I'm making, that's a memory, so my research is trying to dampen those memories. You couldn't really say that it's related to addiction at all, other than to say "Hey, maybe if we can do that in rats, maybe the same pharmacotherapies might have the same effect in humans."

NCN: Yeah, well, I was going to mention that.

ALEX: At some point, in about seventy-four years. [laughter]

Even though the research he was doing was designed specifically to investigate preferences for drugs that humans abuse, Alex resisted the idea that his model was a close analog for human addiction or that it had immediate relevance for human treatment. In keeping with the translational rationale for animal behavior genetics research, he acknowledged that his work might generate some links between the drugs that change his rats' behavior and drugs that could treat human addicts. And yet, he portrayed translation as something that would take place in the distant future—seventy-four years into the future, to be precise.

Although the way that Coast researchers talked about their work as both intimately linked to human illness and very basic seems contradictory, both elements were central to working one's way out of the deadlock of the complexity crisis and toward a promising future for the field. Stretching out the time line of the field's promissory future altered the character of their knowledge production present—it allowed researchers at Coast to live and work in a period of extended uncertainty while still seeing themselves as engaged in a long-term scientific project of generating stable, clinically useful facts.

This characterization of animal behavior genetics as operating on a long time horizon was one that was remarkably consistent among animal behavior geneticists, even outside of the close-knit circle of Coast University researchers. While not all practitioners placed the clinical payoff from their work several scientific lifetimes into the future, they typically worked to extend, rather than compress, the time lines that emerged in the course of conversations or interviews. Dr. Thomas Schmidt, a German researcher I met at a meeting of the International Behavioral and Neural Genetics Society, talked about his mouse models of anxiety in a way that departed noticeably from the speech conventions I had become accustomed to at Coast. Rather than engaging me in complexity talk, Dr. Schmidt talked freely about "anxiety genes" that generated strong predispositions to psychiatric disorders. But when it came to the future of clinical treatment, he complicated his prior characterization by emphasizing that it was likely that there were many such genes, each of which contributed a small amount to an overall genetic predisposition:

TS: That might explain why we don't have the wow, the breakthrough in a sense that somebody discovered the magic bullet. There is none. And that's the problem. So it is step-by-step, this is a 5 percent contributor, OK, it's accepted by the scientific community, so let's go to the next. Then if we found twenty [of these] 5 percent contributors, then we are close to 100 percent, and then we can try to characterize patients in a sense that we can design a cocktail that is most promising for this particular person.

NCN: So it will probably be more of an incremental breakthrough where you'll slowly find more genes that influence it and maybe drugs that act on those systems.

TS: Yeah, at least in mice. And then we still have to look for homologies in the clinic. This is another long way to go, but there's agreement that we should start with mice.

Dr. Schmidt converted an implicit vision of a future where breakthroughs in identifying "anxiety genes" were possible into one where research would proceed in small increments, gradually identifying and verifying the many genes that contributed to anxiety disorders. When I reflected this vision back to him for confirmation, he inserted even more distance into my time line by pointing out that there was an additional translational step that needed to happen between the mouse and the human even after promising gene candidates were identified.

In grant applications, the things that researchers at Coast tended to promise were waypoints along the translational pathway, things that would still need to be translated from mouse to human to be of clinical value. Many of the goals that they outlined were in the realm of test development, such as establishing new ways to measure behavior, or developing and maintaining lines of mice that would be used by the broader research community. Other specific aims focused on characterizing the relationship between the behaviors they had defined and particular genes, neural circuits, or other behaviors. For example, researchers might propose experiments to see whether mice susceptible to seizures when withdrawing from alcohol also experienced increased pain sensitivity, which are both symptoms seen in humans going through withdrawal. The ultimate goal of this research, as Dr. Smith put it in one of his grant proposals, was to "identify the genes responsible for increased risk for and protection against the several different symptoms of alcohol dependence."

Researchers did not give specific time lines for these long-term goals in their grant proposals, but they created a vision of a future similar to the one Dr. Schmidt described, where research would proceed through a series of small, incremental steps. The background and significance section of one of Dr. Smith's grant proposals began, perhaps unsurprisingly, with a section titled "Alcoholism Is Complex." The section described evidence for the polygenic and multigeneic nature of alcohol dependence, and the heterogeneity of the category of alcohol dependence itself. The proposal went on to argue for the need to better characterize alcohol dependence and the genes influencing different aspects of dependence. Human risk flows from the effects of multiple genes, the proposal argued, and at present, there were only a few genes "for which [researchers were] absolutely assured that the polymorphisms play[ed] a role in human alcoholism." Developing new mouse lines and tests would help researchers identify additional genes or neural circuits that were important to alcoholism, which would in turn help other researchers better characterize clinical risk and develop treatments. By assisting in that process of characterizing a complex disorder, Dr. Smith argued that mouse research provided "a powerful foundation for translational research that brings us closer to identifying genes important in determining liability toward the development of alcoholism, and developing gene and drug therapies to enhance prevention and treatment."

This all sounds like fairly standard fare for translational research, but this particular proposal was a renewal of a grant that Dr. Smith had held for thirty-four years, since the very beginning of his scientific career. While Dr. Smith had undeniably had success in establishing new tools and methods in alcohol research (one of the tables in this grant proposal listed nearly a hundred researchers who had used his mouse lines), I was surprised that he could still make the case three decades later that more mouse lines and behavioral tests were needed in alcohol research. Moreover, the proposal was littered with allusions to past experimental problems and failures. Referencing Dr. Smith's failed attempt to classify existing intoxication tests that I described in chapter 1, the grant proposal pointed out that "experience [had] taught [them] . . . that [their] ability to intuit which tasks would be genetically correlated based on these conceptual notions was limited to nil." The types of studies that would be required to unravel the relationship between existing tests, the proposal continued, would be so time consuming that it would be unlikely they would "achieve [their] overall goal in a lifetime."

I questioned Dr. Smith about how he could possibly secure funding while making such seemingly pessimistic claims about the likely outcomes of his work, but his grant track record seemed to speak for itself. Dr. Smith was highly successful in obtaining support from federal granting agencies for his work, perhaps best evidenced by the half a dozen staff he employed, in addition to grad students and postdocs, to carry out all of his funded projects. And as in other aspects of his scientific life, Dr. Smith took pride in the cautious promises he made in his grant proposals. In response to an inquiry of mine about whether he made more forceful claims in funding applications than he did in other venues, he e-mailed me half a dozen of his most recent proposals, and encouraged me to read them for myself. He told me that if we went back far enough in his grant history, I would find a few "smoking guns," such as a grant application he wrote in the late 1970s where he had promised to develop an animal model of alcoholism. But since that application, he told me, he had been careful to emphasize that the aim of his research was to develop *partial* models of alcoholism. Given the kind of bombastic overclaiming that seems to be standard in fund-raising venues, I was surprised that this particular claim was the one that he remembered and regretted. If anything, I expected to hear about naïve, youthful promises to produce breakthrough treatments, not promises to develop an animal model of alcoholism.

To be sure, pessimism is not absent from mainstream visions of the future of genome science. Richard Tutton (2011) has noted that hopeful futures go hand in hand with the articulation of much less desirable possibilities. The disclosures embedded in biotechnology companies' "forward-looking statements," for example, paint a vision of a future filled with uncertainties and pitfalls where hopes may fail to be realized. But Tutton points out that not all promissory statements are created equal. "These statements delineate futures that are hoped to be unrealized," he argues, "futures that are contingent and uncertain, and are precisely not envisioned as performative" (Tutton 2011, 419). While these kinds of legally mandated statements are intended to serve as a counterbalance to the optimistic speculations of companies seeking investors, Tutton argues that speculations about "futures to be avoided" could also serve to put the more optimistic futures on firmer footing by suggesting that companies have already accounted for potential pitfalls.

At Coast University, in contrast, the multiple gaps and uncertainties that are reduced to boilerplate text in forward-looking statements were fleshed out by retelling stories about overpromising, controversy, and failures to replicate. Unfettered optimism and the incautious claiming that comes with it became the future to be avoided, informed by past researchers' overly ambitious estimations of their capacity to unravel the complexity of behavior using new techniques. Researchers managed what they perceived to be overly confident and socially irresponsible knowledge claims by expanding the field's temporal horizons back out again, pushing clinical scenarios into the far distant future and introducing potential complications that made it seem unlikely that clinical applications would be coming any time soon.

Extending the time horizon of one's research efforts and deflating hopes for the present may not seem at first to be a recipe for successfully ensuring the legitimacy and power of ones' scientific enterprise, but this narrative fits neatly into the landscape of contemporary genetics. Coast researchers drew on both the hope of a future where genomics could improve human health as well as the accusations of "hype" and overpromising that haunted other genomics fields and more entrepreneurial practitioners. As self-described basic researchers whose work connected to a valued but distant future, they created a space in which the promise of their near-term work lay in building up the epistemic foundations of the field rather than in producing specific, enduring facts. This rhetorical formula, as Arribas-Ayllon, Bartlett, and Featherstone (2010) have similarly argued, constructed optimistic but moderate expectations about what the field was likely to accomplish.

Coast researchers' portrayal of themselves as cautious scientists—ones who were concerned with crafting statements that reflected the complex reality of behavior over simple stories that would bring them attention—also resembles long-standing tropes such as Robert Merton's (1973 [1942]) functionalist description of science. Merton's norms of science articulated an aspirational vision of epistemological modesty, one where scientific practitioners disavowed any investment in the outcome of particular experiments or desire to profit from them, and instead subjected themselves and others to regular skeptical scrutiny. The ideals of scientific behavior embedded in Coast researchers' past and future stories drew on this image of science and the scientist. Through retelling stories in which they as protagonists applied the brakes on "speed genomics" (Fortun 1999) by arguing against the hopeful findings of new molecular studies or even undermining their own published work, researchers at Coast performed the norms of disinterestedness and organized skepticism that still hold so much cultural value—perhaps especially so in a historical moment when concerns run high about how commercial interests might be undermining the integrity of biomedical research.

CONCLUSION

In the course of my hours-long interview with Dr. Clark, I mentioned a news story I had seen a few days before, titled "Ever-Happy Mice May Hold Key to New Treatment of Depression" (McGill University 2006) and describing a study on depression recently published in *Nature Neuroscience* (Heurteaux et al. 2006). The researchers had developed a knockout mouse line missing a gene that coded for an ion channel involved in regulating serotonin in the brain. They tested the mice in several behavioral models of depression and found that the knockout mice behaved similarly to mice that had been treated with antidepressant drugs. The news article speculated that this "permanently cheerful" mouse that was "resistant to depression" could lead to the development of a new generation of antidepressant drugs. Dr. Clark leaned back in his chair as I summarized the article for him, looking up at the ceiling and shaking his head. "I reel from that stuff, Nicole," he said when I'd finished. "I think it's very harmful."

The historical contingencies that have shaped the field of animal behavior genetics and the ways in which practitioners retold this history created a culture in which scientific claims making became a matter of intense concern. In the heterogeneous field of animal behavior genetics, knockout studies such as the one I described to Dr. Clark reactivated disciplinary fault lines between those who approached the field from the direction of molecular biology and those who saw themselves as specialists in animal behavior—which in turn mapped onto differences in understandings of gene action and styles of claims making. Making statements about the nature of depression and the future of treatment based on studies of a single missing gene went against Coast researchers' assumptions about the complex nature of behavior. But more than that, such simplistic stories also had the potential, in their eyes, to damage the credibility of the field and provoke social discord. Researchers at Coast used these disciplinary differences as foils for justifying particular methodological stances and articulating key scientific commitments. Retelling stories about the field's history was also a way for researchers to socialize new members of the field and draw boundaries around their scientific community.

Moderating claims, preserving uncertainty, focusing on the epistemic foundations of the field, and pushing production of specific facts or clinical applications far into the future—all these were techniques researchers at Coast used to mitigate what they saw as dangerous tendencies toward overclaiming in their field, and perhaps in genomics more generally. In doing so, they crafted a unique narrative about their field and what it could deliver, one that drew on both genohype and its criticisms. The familiar promise of using genomics research to transform human health framed their stories, but it was remixed with warnings about the dangers of moving too quickly and claiming too much. Together, Coast researchers' pasts and futures helped to create and sustain a culture where animal behavior genetics facts could seem to be permanently under construction. The question of how researchers operated in this unfinished space is what we will turn to next.

CHAPTER 3

Building Epistemic Scaffolds for Modeling Work

A mouse poked its head around the edge of a plastic wall. Whiskers twitching, it strained its neck around the wall, sniffing and darting its head from one side of the opening in the wall to the other. A single lamp was mounted above the plastic maze, illuminating it like a stage in the center of the otherwise dark laboratory room. The room was silent except for the hum of the computer fan and the soft scratching sounds of other mice in their cages. I sat with a technician in the corner of the room, and we observed the mouse's progress through the maze on the computer screen. The mouse moved in fits and starts, stretching its body forward and then jerking back, taking a few quick steps ahead and then freezing. A video camera mounted above the maze silently recorded these movements. As the mouse passed through the opening in the wall and out onto a plastic ledge, the video tracking software counted one "open arm entry." On the screen, the image of the open plastic ledge lit up in bright green as the mouse explored this area. The mouse advanced slowly. It paused again after a few steps, dipping its head over the edges of the plastic balcony. Then suddenly, the mouse swung its body around, tail flicking out over the side of the narrow ledge, and scampered back through the opening in the wall. The green highlighting disappeared from the computer screen as the mouse retreated into the enclosed areas of the maze. We waited. As the end of the mouse's allotted fifteen minutes in the maze approached, the technician got up. When the timer sounded, she deftly plucked the mouse from the maze by its tail. She wasted no time in wiping down the maze with a cleaning solution, restarting the timer, and retrieving the next mouse from its cage.

The experiment that was taking place in this dimly lit laboratory room is

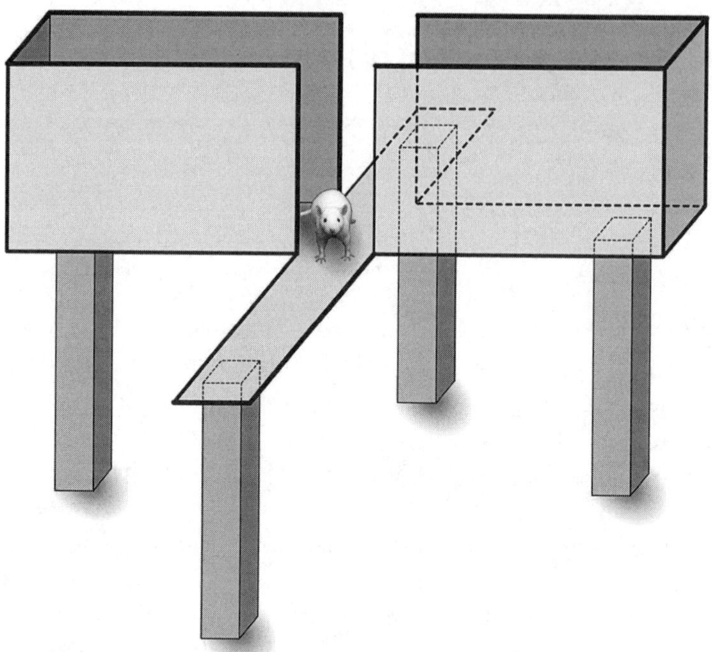

FIGURE 3. Schematic of an elevated plus maze. Illustration adapted from original work by samuel-john.de, licensed under the Creative Commons Attribution-ShareAlike 3.0 Unported License (http://creativecommons.org/licenses/by-sa/3.0).

known as an "elevated plus maze" experiment. As the name of the test suggests, it involves a simple maze in the shape of a plus sign that is elevated about half a meter off of the floor (figure 3). Two arms of the plus sign are surrounded by high walls, which are called the "closed arms" of the maze. The other two arms are narrow, "open" platforms without any enclosures. At the beginning of a testing session, the researcher places a mouse in the center of the maze where the open and closed arms meet, and the mouse is left to explore the maze for a short period (typically between five and fifteen minutes). Researchers track how much time the mouse spends in different areas of the maze using a video camera mounted above the maze, and the mouse's behavior is then scored manually by a trained observer watching the video or automatically by video tracking software. Measurements such as the number of times the mouse enters the open arms, and the percentage of time it spends there, form the basis for assessing the anxiety level of the mouse. The more time that the mouse spends in the open, unprotected areas of the maze, the less anxious it is said to be.

The maze was first described in 1985 as a "novel test for the selective identification of anxiolytic and anxiogenic drug effects in the rat" (Pellow et al. 1985, 149). It was quickly adapted for use with mice by Richard Lister, an intramural researcher at the National Institute for Alcohol Abuse and Alcoholism (NIAAA), who published the first article in 1987 describing mouse data from the test (Lister 1987). Behavioral tests such as these, which practitioners also interchangeably describe as behavioral "models" or "paradigms," are one of the basic working units of animal behavior genetics research. They are well-characterized experimental tools where the equipment and testing protocols have been described and to some degree standardized, work to validate the model has been conducted, and the baseline behavior of common mouse strains is already known. Each model is linked to a facet of a human behavioral disorder, from broad categories of behavior such as anxiety and depression to narrowly defined traits such as "novelty-seeking behavior." These existing tests are then used to set up new experiments. Researchers today use the elevated plus maze to run a wide variety of experiments on anxiety, from large-scale screens of potential anxiolytic (anxiety-relieving) drugs to experiments investigating whether particular genetic manipulations alter anxiety. Researchers often refer to it as the "gold standard" of anxiety tests, and it would not be an exaggeration to say that tens of thousands, perhaps even hundreds of thousands, of mice have been sent through this maze over the decades in the name of research.[1]

Despite the test's popularity, it is far from immediately obvious why observing a mouse in a plus-shaped maze is a credible means of generating knowledge about anxiety. In my experience, observers unfamiliar with animal behavior genetics often reacted with skepticism or incredulity at the idea that the mouse's behavior in this maze was related to human anxiety in any meaningful way. At one academic conference where I described the elevated plus maze, the question and answer session quickly veered away from my arguments about knowledge production and toward questions about the test's validity. One audience member commented that she did not find this test to be a plausible model for human anxiety because a human faced with this situation would instinctually do the opposite of what the mouse does—walk out onto an open arm and have a good look around to assess the situation. Another audience member approached me after the session to say that she thought the first comment was absolutely wrong, and that she would stay in the closed arms of the maze. A commenter at another conference suggested to me that perhaps the mouse is not afraid of open spaces but afraid of heights and that researchers should reconsider the wisdom of building the maze floors and walls out of clear plastic.

Although the terms of the debate are different inside the animal behavior genetics community, the validity of the test is no less a topic of discussion. Even though it has been in use for decades, there are still lively conversations about the best way to conduct and interpret elevated plus maze experiments. New methodological studies of the maze continue to appear in the literature, exploring pragmatic questions such as whether building mazes out of plastic or metal affects experimental outcomes (Hagenbuch, Feldon, and Yee 2006), or suggesting improvements to the original maze design (Fraser et al. 2010). Some researchers have proposed a new design, the "elevated zero maze," with the open and closed platforms linked together in a continuous circle to eliminate the problem of how to interpret the behavior of mice who sit at the intersection of two arms in the plus-shaped maze (Shepherd et al. 1994). Theoretical discussions also continue about whether the test really measures what it is supposed to. Several research groups, for example, have undertaken statistical analyses to examine whether the measurements taken in the maze can really be said to measure "anxiety" or whether they are confounded by other factors, such as the activity levels of the mice (Milner and Crabbe 2008; Wall and Messier 2000).

To better understand this abundance of methodological discussion, this chapter develops the metaphor of building "epistemic scaffolds" to support knowledge production work. This phrase draws on long-standing analogies in science and technology studies (STS) between construction work and scientific work, and uses them to examine the extrafactual work researchers engage in to establish mouse models as credible means of producing knowledge. With expectations for firm findings pushed far into the future, this methodological work was especially visible and valued, with many of the leaders of the animal behavior genetics community engaged in the work of establishing the methodological foundations that they hoped would allow for future progress in understanding the genetics of behavior. A few researchers at Coast had made something of a professional specialty of model development, devoting a substantial proportion of their resources to experiments and publications aimed at creating new models or refining existing ones. But methodological refinement also took place in much more informal ways in the laboratories at Coast, such as in exchanges where new researchers learned to refer to the elevated plus maze as a test of "anxiety-like behavior" and scan the methods sections of published papers for certain key phrases. The metaphor of scaffold work draws together these seemingly unrelated activities and discussions into a larger process of negotiating the epistemic foundations of the field.

EPISTEMIC SCAFFOLDS SUPPORTING ANIMAL MODELS OF HUMAN DISORDERS

The scaffold is a useful object for thinking about how researchers assemble the rationale for an animal model because it functions as a support structure and platform for doing work, and because it is a transient structure that can be modified, reconfigured, and adjusted to different heights.[2] The purpose of building a scaffold at a construction site is to create a surface from which to work on more permanent structures, just as model validation work is aimed at establishing protocols that will be used to generate supposedly more enduring genetic or neuroscientific findings. When the permanent structures are completed, the scaffold is ideally dismantled, leaving no trace of its role in the production of the finished product. But as we will see in the case of the elevated plus maze, epistemic scaffolds (like their real-life counterparts at construction sites) often end up becoming permanently provisional structures: they are built for particular pragmatic purposes, but the work that they are needed for takes longer and is more complicated than expected, turning these supposedly transient structures into semipermanent features of the scientific landscape.

The design of a scaffold and the materials used to construct it affect how stable and useful it will be for doing work. While construction workers or researchers might build a scaffold with whatever is at hand in whatever manner they choose, there are some accepted techniques for building a stable platform. In animal behavior genetics, using pharmacological testing to link animal experiments to human behavior was a particularly effective way of validating a mouse model, because this evidentiary link weathered critiques of anthropomorphism especially well. Basing a model on "face validity," where the resemblance between animal and human behavior justifies the use of a particular test, offered less support because it was easily critiqued as too subjective a link. All scaffolds become shakier and require more support as they are built up to greater heights. Particularly in animal behavior genetics, where even the permanent factual edifices that researchers were working on seemed unstable, practitioners frequently voiced concerns about whether the field's epistemic scaffolds provided safe platforms for doing work. In the absence of an outside party who could certify that particular models were strong enough to bear the epistemic weight that they were expected to carry, different researchers sometimes made quite different assessments about how high scaffolds should be built and what kind of work could safely be done on them.

My use of the scaffold metaphor takes inspiration from ethnographic descriptions of how scientists construct and dispute knowledge claims, especially Trevor Pinch's (1985) and Bruno Latour's (1987) descriptions of how claims are built up from specific observations. Pinch (1985) describes the relationship between a knowledge claim and the observational data it is based on in terms of "externality." In Pinch's terms, a scientist's claim that she recorded "splodges on a graph" has a low degree of externality, while the claim that her experimental traces are evidence of a particular ion has a higher degree of externality, and the claim that this ion indicates the presence of solar neutrinos has the highest degree of externality. Claims with low externality tend to be safer but also more modest in what they propose to contribute to the field, while claims with high externality are bolder but also more likely to draw criticism from colleagues. Latour (1987) describes a similar process of "fact building" where scientists "stack" claims on top of each other to create claims with higher degrees of "induction" so that data from three hamsters' kidneys are transformed into claims about kidney structures in mammals more generally.

Building an epistemic scaffold similarly involves a process of building up a structure of evidence and arguments to make claims about a behavioral test. What is "stacked" in an epistemic scaffold, however, is not a series of increasingly general claims about a particular observation but rather a series of increasingly risky claims about a model's knowledge production capacities. It is worth emphasizing that I speak about the riskiness of the claims embedded in epistemic scaffolds, which is related to but not synonymous with their degree of generality or abstractness. Claims about either facts or models that are broad in scope may in general be more difficult to support than narrow ones, but whether this is true in any particular case depends on the context in which that claim is made and the audiences that evaluate it. For example, there is no intrinsic reason why a genetic factor affecting behavior should be considered any more abstract than an environmental one. And yet, practitioners are likely to see the claim that a test can detect a specific gene's contribution to anxiety as more risky than the claim that a test reveals how a technician's handling alters anxiety levels. Funding agencies or pharmaceutical companies are similarly more likely to be interested in (and skeptical of) claims about tools for producing genetic information because of the social and economic value of genetic data.

Unlike claims about particular observations that are typically produced by an individual researcher or a laboratory group, behavioral models are shared entities used by many practitioners in the scientific community. This is consequential for understanding differences in how factual claims and scaffold claims are developed and debated. Analysts' descriptions of how factual claims are built up are typically framed in terms of an adversarial context. Latour (1987, 51), for example, describes the process of claims making in self-interested terms: he argues that successful scientists will attempt to "prove as much as [they] can with as little as [they] can considering the circumstances." Whether these claims survive, in his view, depends on the intensity of competition with other scientists in the field. In the case that I describe, however, those who are participating in breaking down the epistemic scaffolding of particular models and advocating for more conservative views of their utility are not rival laboratories, but are often researchers who are themselves users (or even designers) of these models. In discussing collectively used models, theories, or scientific objects, other professional concerns and interests aside from interpersonal competition come into play, such as competition between scientific fields or the credibility of a particular field in society at large (Abbott 1988). Discussions about the validity and utility of animal models are thus negotiations that take place at the level of the field, and with an eye toward how these methods will be regarded outside of the field. These kinds of dynamics are particularly important for understanding claims making in behavior genetics because (as I explored in chapter 2) practitioners were especially concerned with the potential social consequences of the claims they made. Researchers wanted to be circumspect about the capacities of animal models as knowledge production tools, while at the same time not being so cautious that the field was left with no tools at all. This tension animated many of the seemingly contradictory arguments researchers made, where they built up the arguments and evidence supporting the use of a mouse in a plastic maze as a credible means of gaining insight into a human behavioral disorder even as they simultaneously pointed out the radical limitations of such an endeavor.

BUILDING UP SUPPORT FOR THE ELEVATED PLUS MAZE

To examine how researchers made the case that the elevated plus maze was a useful model for human anxiety, let us begin with a description of the test from the website of Panlab Harvard Apparatus, a manufacturer of behavioral test equipment. This short description succinctly outlines some of the main technical arguments researchers use to support the elevated plus maze as a model of anxiety. They write: "The elevated plus maze is a widely used animal model of anxiety that is based on two conflicting innate tendencies: exploring a novel environment and avoiding elevated and open situations constituting situations of predator risk. . . . When placed into this apparatus, naïve mice and rats will, by nature, tend to explore the open arms less due to their natural fear of heights and open spaces. In this context, anxiolytics generally increase the time spent exploring the open arms and anxiogenics have the opposite effect, increasing time spent in the closed arms" (Panlab 2010). The two lines of argument embedded in this brief description each call back differently to the natural/artificial properties of the mouse as model.

The first argument, which I call the *pharmacological argument*, is based on evidence of the effects that anxiolytic (anxiety-relieving) or anxiogenic (anxiety-inducing) drugs have on the behavior of mice in the maze. This argument emphasizes the controlled nature of the experiment and the ways in which the mouse has been altered to serve the needs of the experiment, portraying the mouse as a kind of lively biological detector for drug effects. What the mouse is experiencing and whether it resembles human anxiety in any way is largely irrelevant; what matters is that mice behave in a certain way when they are given anxiolytics, and that this behavioral change accurately predicts which drugs will be effective for relieving anxiety in humans. The argument I call the *ethological argument*, in contrast, draws more deeply from natural history of the mouse to make a connection to human anxiety.[3] This line of reasoning employs knowledge about the "innate tendencies" of mice to explore some spaces and avoid others to support the idea that the test is a good model for anxiety. Although some researchers argue that mice do not experience anxiety disorders in exactly the same way that humans do, the ethological explanation assumes some sort of evolutionary relationship between the behaviors and biological responses that mice exhibit and what humans identify as anxiety.

The pharmacological argument is based on a series of experiments in which researchers gave mice drugs known to increase or decrease anxiety in humans and then tested them in the maze. In his original article, Lister (1987) reported that when he administered chlordiazepoxide (a drug in the benzodiazepine class, better known by its brand name Librium) to mice, they spent more time in the open arms of the maze. Conversely, when Lister gave mice caffeine, which increases anxiety in humans, they spent less time in the open arms. Lister used these observations to make an argumentative link between mouse behavior in the maze and human anxiety. The claims that Librium relieves anxiety while caffeine provokes it in humans are widely accepted in the scientific community, and linking this established information to corresponding changes in the behavior of the mice in the maze forms the base of this section of the scaffold supporting the elevated plus maze (figure 4).

On its own, this link between mouse research and human experiences may not be that useful. The claim that the elevated plus maze can be used to detect the effects of Librium or caffeine is a safe claim with a low degree

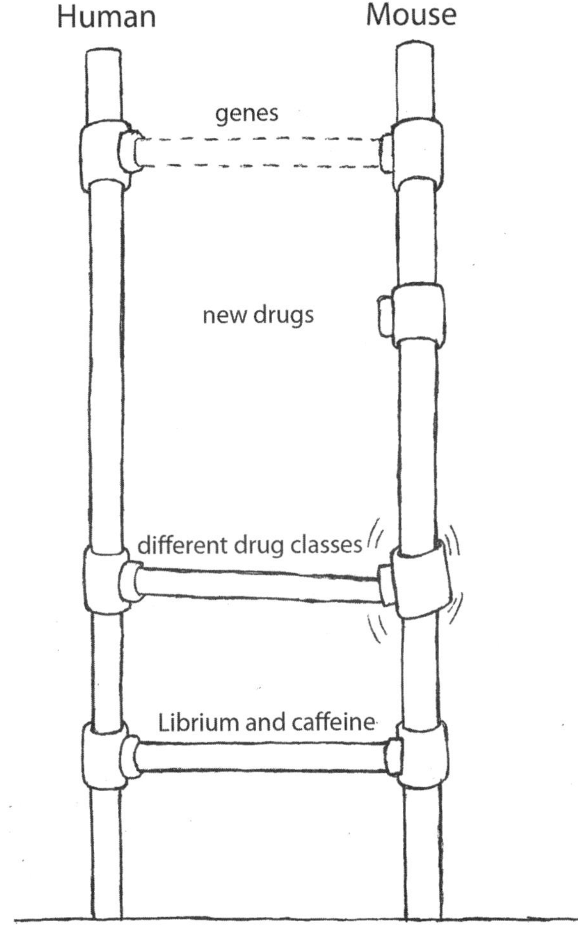

FIGURE 4. The pharmacological argument supporting the elevated plus maze. Illustration drawn by the author.

of "externality"—it is easy to defend, but it does not make the elevated plus maze appear particularly useful for doing research work. This link, however, forms a foundation from which researchers can build upward. Based on this observation, researchers have argued that the test can also be used to investigate the effects of other anxiolytic or anxiogenic drugs; or, even more broadly, to screen new pharmacological agents for antianxiety activity (Walf and Frye 2007). Going even further, some have argued that this test can not only find more drugs that affect anxiety but can also identify other kinds of factors that affect anxiety. For example, researchers have employed the test to investigate whether knocking out particular genes changes a mouse's behavioral patterns in a way similar to administering an anxiolytic drug. This expansion of the test's knowledge production capacities is evident in the way that researchers have begun to describe it in the literature as a tool that permits both "a rapid screening of anxiety modulating drugs [and] mouse genotypes" (Bourin et al. 2007, 570).

To most researchers, claims that the elevated plus maze can be used to find new drugs or genes that affect anxiety in humans required some kind of additional support. The elevated plus maze may have worked well for detecting the effects of the particular drugs that Lister studied, some may argue, but what if those patterns did not hold true for other drugs? Researchers accordingly set to work on the epistemic scaffold of the test, conducting further experiments to study the behavioral effects of administering anxiolytic drugs from other pharmacological classes to mice in the maze. The mixed results of these experiments have provided material both for those wanting to make riskier claims about the test's knowledge production capacities and for those wanting to restrict its use. Studies of other drugs showed that not all followed the same pattern as those studied by Lister: while drugs in the benzodiazepine class reliably changed the behavior of mice in the maze, other drugs, such as buspirone (an anxiolytic chemically unrelated to the benzodiazepines) and fluoxetine (a selective serotonin reuptake inhibitor commonly described as an antidepressant but that also has antianxiety effects), produced inconsistent results. Some studies showed that these drugs had no effect on, or even slightly decreased, the time that mice spent in the open arms (Kurt, Arik, and Çelik 2000; Moser 1989; Silva, Alves, and Santarem 1999). Incorporating evidence from these additional experiments into the elevated plus maze's epistemic scaffolding thus generated some potentially weak links. Those who wanted to extend claims about the capacity of the test could use that evidence to argue that the correlations between mouse and human behavior held true for many several drug classes, but critics could also quickly point to the exceptions to those general patterns. One review article of animal models of anxiety made just this criticism, arguing that because anxiety models such as the elevated plus maze had been developed and validated primarily with benzodiazepines, "their sensibility on drugs acting on other system remains questionable" (Bourin et al. 2007, 573). Some researchers have attempted to repair this particular weakness in the scaffold by arguing that the inconsistent findings with different classes of drugs may be due to variations in the way that other researchers have conducted their experiments, and that further standardization of the test protocol might eliminate such inconsistencies (Hogg 1996).[4]

The further the claims about the knowledge production capacities of the elevated plus maze are built up, the shakier the epistemic scaffold becomes, as claims appear more risky and open to potential criticisms. The pharmacological scaffold is particularly shaky in the upper regions, where the strength of the claim that the mouse model can be used to discover new drugs that might work in humans has been heavily contested. Practitioners have also expressed differing opinions about whether the test is suitable for producing knowledge about genetic components of anxiety. A particularly strong critique of the elevated plus maze found in one methodological article made both of these arguments. Dawson and Tricklebank (1995, 36) wrote that the test "has yet to make a major contribution to the discovery of a novel anxiolytic or to further our understanding of either the psychological or physiological basis of anxiety or its relief," and that therefore "it is difficult to justify its use as anything other than a preliminary screen as a prelude to more robust testing in animal models of anxiety."

The second line of argumentation, the ethological argument, draws on a different body of knowledge coming from studies of the behavior of animals in their natural habitats. This argument contends that the maze is a good test for anxiety because it re-creates a type of conflict that mice experience in their natural environments, one that also mirrors conflicts humans experience. Studies of mouse behavior in natural habitats have shown that they are inclined to explore new spaces to find sources of food but also wary of exposed areas that might leave them vulnerable to predators. The instinct to explore a new space is thus pitted against the instinct to avoid a dangerous place, creating what psychologists call an "approach/avoidance conflict" (a concept first articulated in human psychological research by Kurt Lewin in a 1935 book). When the balance between these two instincts tips too far in favor of caution, mice will avoid the open spaces almost entirely and remain instead in protected areas, a pattern of behavior that researchers have argued resembles the pathological avoidance of stressful situations seen in human anxiety suf-

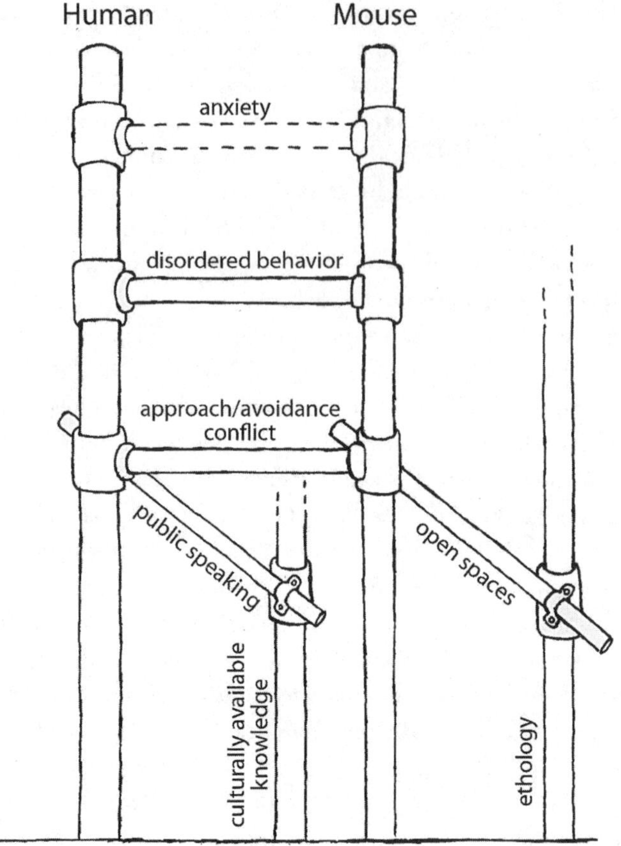

FIGURE 5. The ethological argument supporting the elevated plus maze. Illustration drawn by the author.

ferers. The claim that the behavior of both mice in the elevated plus maze and humans in anxiety-provoking situations can be understood as a form of approach/avoidance conflict forms a foundational link between the mouse and the human in this section of the epistemic scaffold (figure 5).

The kinds of evidence that the ethological argument employs in making the case for the epistemic capacities of the test are quite different than those of the pharmacological argument. In contrast to the controlled drug experiments that are foundational to the pharmacological explanation, here observations of the natural behaviors of research animals are the core of the argument. What counts as a "natural" environment or a "natural" behavior for rodents that have been bred for generations in laboratories, however, is open to question. Lister (1987) noted that the researchers who originally proposed the rat version of the test were inspired by observations of rat behavior made in laboratory environments rather than the field. When designing the new test, Sharon Pellow and colleagues (1985) drew specifically on observations made by psychologists in the 1950s that rats exploring a maze generally preferred to stay in narrow, dark hallways rather than in open or brightly lit spaces. But researchers have also made arguments that more closely resemble classical ethology, explaining how particular behaviors mice exhibit in the maze make sense in light of the mouse's evolutionary history and might have contributed to the animal's chances of survival and reproduction outside of the laboratory. Another methodological article on the elevated plus maze took this approach, describing the logic of the test by stating that "mice are prey for many other larger animals, which may underlie their natural tendency to avoid open and, thus, unprotected, spaces (and, to a lesser extent, heights)" (Walf and Frye 2009, 228). No matter how the "natural" environment of the mouse is argumentatively constituted, though, what unites this family of arguments is that observations of mouse behavior under natural conditions are enrolled in justifications for the test design.

That the behaviors of two distantly related mammalian species can both be characterized as forms of an innate approach/avoidance conflict is already a fairly risky claim. Some researchers have questioned whether the behavior of mice in the maze is really analogous to the human situation, or whether anxiety in humans can be accurately characterized using this psychological concept. Some have claimed in the literature that the test has solid "face validity" (Walf and Frye 2007), others have argued that the resemblance between the test and the human disorder is "questionable" (Haller and Alicki 2012, 59). When recapitulating the ethological case for the validity of elevated plus maze, then, practitioners often recruited additional evidence to support this foundational link. In Crawley's (2007, 230) textbook on behavioral testing in mice, she elaborated on the approach/avoidance concept by providing an example of an anxiety-provoking situation that researchers might face and analogizing it to the predicament faced by mice in the maze: "You want to tell the world about your exciting research results, but have fears about the audience's response when you walk up to the podium to give your talk. A mouse may want to explore a new environment to find food, but may fear venturing out into the open where it is an easy target for predators." This comparison draws on both culturally available knowledge about public speaking and on ethological knowledge about mouse behavior to support the argument that aspects of both mouse and human experience can be characterized as approach/avoidance conflicts.

From this foundational human–mouse link, researchers have argued that the elevated plus maze can be used not only to model conflicts between approaching and avoiding in particular situations, but also to investigate an extreme version of this behavior that resembles what clinicians would diagnose in humans as an anxiety disorder—the difference between what psychologists call "state" anxiety that is a temporary response to a specific situation, versus "trait" anxiety that refers to durable differences between individuals in how they perceive and respond to threats. This claim, of course, can be contested: some researchers have argued that the elevated plus maze is limited in its utility because a test of such short duration can only be said to model the "state" anxiety that particular situations induce in all humans, rather than the continued anxiety experienced by anxiety disorder sufferers (Haller and Alicki 2012). Others have argued that because humans with pathological anxiety also respond more strongly to situations that provoke anxiety in normal people, the test can be said to provide information on both nonpathological, situation-specific anxiety and disordered behavior (Ramos 2008, 496). An even riskier move would be to claim that the elevated plus maze simply measures "anxiety," without further qualification. Building the scaffold up to that height is almost certain to invite criticism, and perhaps consequently such claims rarely appear in the published literature. Where they do appear, though, is in everyday conversations in the laboratory, as we will examine next.

DYNAMICS OF EPISTEMIC SCAFFOLDS

Even from this brief examination of the published literature, the extent to which animal behavior geneticists actively discussed, debated, and modified the epistemic underpinnings of their animal models is already evident. The arguments and evidence in support of using the elevated plus maze as a model for anxiety were not established once and then held stable but continued to be subject to ongoing revisions and restatements, both large and small. Practitioners actively modified the epistemic scaffolds of their models by introducing new evidence or criticizing particular existing claims. By acting on epistemic scaffolds in these dynamic ways, individual practitioners and communities of practice developed specific and local understandings about the value and limitations of particular behavioral tests.

One way that I observed practitioners reconfiguring epistemic scaffolds, both in formal and informal settings, was by generating novel combinations of the various existing arguments that supported particular tests. Although for the sake of clarity I have presented the pharmaceutical and ethological arguments as distinct arguments here, in practice, researchers drew fluidly from these and other arguments when talking about the validity and utility of animal models. These dynamics were especially evident in one workshop session where I observed a researcher introducing an audience new to behavioral testing to the available rodent tests of anxiety. When she reached the part of the agenda dealing with the elevated plus maze, she started by giving a general description of the ethological justification for the model. As a behaviorally trained researcher this evidently seemed to her to be the most logical place to start in describing the theoretical underpinnings of the model, but the audience of new practitioners seemed unconvinced. After fielding several skeptical questions about whether a mouse's avoidance of the open arms of the maze was really analogous to human anxiety, she spontaneously switched over to the pharmacological argument and explained that drugs known to be effective in human anxiety also modified these behavioral patterns. This answer seemed to satisfy her questioners, and she moved on to discussing the practical aspects of running elevated plus maze tests.

Instances such as these demonstrate how connecting several arguments to support the utility of a particular model can give added strength and stability to epistemic scaffolds by creating a dense framework of arguments that is more difficult to break down than a single argument on its own. One of the vulnerabilities of the ethological argument is that it is based on less quantifiable evidence and therefore seemed less "scientific" to some audiences. The experimental evidence marshaled by the pharmacological argument provided a useful corrective to this weakness. Conversely, researchers argued that tests validated with pharmacological information alone could reach mistaken conclusions (e.g., a drug might be mistaken as anxiety relieving or inducing if it simply sedated the mouse so that it could not move out onto the open arms), and have insisted that ethological evidence in favor of animal models is needed to shore up the potential shortcomings of pharmacological evidence.

By linking together several arguments, researchers built up epistemic scaffolds by using the strengths of some arguments to bolster the weaknesses of others. This applied to the flexible ways in which practitioners drew on various arguments for validity of specific tests, but also to the interdependence of mouse models with other more general arguments about similarities between the mouse and the human. In the introductory behavior genetics course at Coast, for example, I was surprised to find that the first class session covered basic evolutionary concepts such as variation within populations, natural selection, and speciation; the syllabus also included a selection from Darwin's *On the Origin of Species* on the reading list. To me, this material seemed like a rather large detour in an introductory behavior genetics course. To Dr. Smith and his colleagues, however, the evidence of evolutionary relationships between mice and humans and the idea that variation in traits could be inherited was foundational to the animal behavior genetics enterprise. It would be difficult to make a case at all for the elevated plus maze were it not for this existing epistemic infrastructure that supported the idea that research with mice can generate useful biomedical knowledge about the human. The scaffolds supporting particular behavioral tests are all supported by these other more general arguments.

Fusing together various arguments into a patchwork of connected scaffolding carries risks as well as benefits, such as the risk that one weak section might compromise the stability of the entire structure. Dr. Amy Lee, a veterinarian I met at a different workshop introducing new practitioners to mouse tests of stress and anxiety, came away from that workshop skeptical about the value of existing animal models of anxiety. In a conversation with me afterward, she reflected on what she had learned and said that she remained unconvinced about the utility of tests like the elevated plus maze because the presenter's ethological arguments simply did not ring true to her:

> I really think a lot of those—a lot of that—maybe anthropomorphism, or maybe—it's just off, it just seemed off to me, you know, a lot of the interpretation of what those tests mean. I think that because behavior is so complex, we can't make those sort of fast leaps like this test is assessing fear. . . . I think that for the [tests] where it has been pharmacologically proven that anxiolytics decrease that behavior, I think there's an argument for it. But I'm still not going to say that the mouse is anxious and we're alleviating it. All I'm saying is that whatever pathways that alleviate anxiety in us exist in the mouse and we're alleviating that as well. Now whether [the mice] interpret that as anxiety or whatever else, I don't know if we can make that leap, scientifically.

Dr. Lee's reaction to her introduction to behavioral testing shows how weaknesses in particular arguments can generate instability in the entire scaffold of the elevated plus maze, and perhaps even in the scaffolding supporting animal behavioral modeling in general. Although she found the pharmacological argument somewhat convincing, her skepticism about the "anthropomorphic" reasoning that she saw in the ethological argument made her doubt the overall validity of the test. If the pharmacological evidence was so convincing, then why were researchers relying on questionable arguments about the mouse's natural behavior to justify the test? If anything, she told me, she thought that the test appeared to measure what she would call "bravado" rather than "anxiety"—a somewhat surprising statement given her objections to the workshop leader's overly anthropomorphic language, which I will return to in a moment.

Dr. Lee's skepticism about the elevated plus maze could be interpreted another way: not as mere skepticism about the validity of the elevated plus maze, but as a different and more conservative view of the kind of knowledge production that the epistemic scaffolding of the test could support. Her comments suggested that she might find it acceptable to use the test for some purposes, such as exploring the shared biological pathways activated by anxiolytic drugs in both humans and animals. But she found it unsuitable for generating the type of knowledge that she needed—data on the baseline stress levels of the mice in her breeding facilities and the impact of any changes that they made to their care practices. Like Dr. Lee, many of the animal behavior geneticists I interacted with were willing to endorse the use of the elevated plus maze for some purposes but not others. Dr. Rachel Jackson, a drug and alcohol researcher at Coast, told me:

> That particular animal model, I'll just tell you some of my bias, was developed for screening drugs that work on anxiety in humans. . . . And for that purpose it's probably a fairly decent test. But when you're looking at, say, withdrawal-induced anxiety, then you have a problem. Mice that are withdrawing from alcohol don't move very much. So how do you interpret their behavior on that maze? You can't. You then don't know if it's behavioral, they're not motivated, you don't know if it's malaise, they just don't feel well. There's no way to interpret those data. So modeling anxiety in mice is a very difficult prospect, and I think it's very important to understand the difference between a screening test for drugs and a behavioral test for an emotive state; [they] are really two separate things.

In Dr. Jackson's opinion, the epistemic scaffolding of the elevated plus maze was robust enough to support its use as a "screening test for drugs" but not as a "behavioral test for an emotive state." Dr. Jackson was pushing back against what she saw as a dangerous trend toward using the test for research questions that it was not designed to answer, an opinion that she told me she often voiced when reviewing articles that put elevated plus maze experiments to new purposes. To illustrate her concern, she described how a researcher might attempt to use the elevated plus maze to measure anxiety levels in mice that were experiencing alcohol withdrawal—a logical experiment to try, given that some human alcoholics report that they experience anxiety during withdrawal, which motivates them to start drinking again, but one which, in her opinion, raised substantial methodological problems. This thought experiment illustrated what she saw as the dangers of using the test for purposes other than drug discovery.

Researchers also attempted to circumscribe the capacities of animal models by including new facts into the scaffolding of those models. In my interview with Dr. Clark, who had extensive experience with running elevated plus maze experiments, he highlighted two details about anxiety as a disease category that were normally elided in descriptions of the elevated plus maze; namely, the fact that "anxiety" in humans is an unstable category whose definition and measurement is contested by clinical researchers, and that many different behaviors fall under this global category. He said, "Are you measuring 'anxiety' or are you tapping into some undefined piece of an anxiety dimension that's hard enough to diagnose in a human? Let's not pretend that this test is a measure of anxiety behavior. No, it's a measure of behavior that's probably somehow related, but let's not start substituting the label 'anxiety' for this behavior."

By drawing attention to these complications, Dr. Clark was attempting to reconfigure the epistemic scaffolding of the elevated plus maze to make the more circumscribed claim that the test was suitable for measuring some as yet unidentified component of anxiety, but certainly not capable of capturing everything that humans might experience as anxiety. Dr. Clark himself took this admonition quite seriously: he told me that he had conducted dozens of experiments with the elevated plus maze, but because he felt he never quite understood what the test was measuring, he could not "bring himself" to publish those experiments. He gestured toward a filing cabinet in his office, where the data from those experiments sat unpublished.

DEVELOPING AND SIGNALING SHARED UNDERSTANDINGS OF EPISTEMIC SCAFFOLDS

By acting on and modifying the epistemic scaffolds supporting the elevated plus maze, animal behavior geneticists constructed multiple, local versions of the "anxiety" that the test measured. The positions that Drs. Lee, Jackson, and Clark articulated on what kinds of knowledge the elevated plus maze test could generate were subtle, making specific and circumscribed connections between mouse experiments and human experience. While researchers had the opportunity to elaborate on such views at length in the classroom or in interviews with curious ethnographers, communicating these specific understandings of a test's capacities in everyday scientific settings more often relied on sets of locally intelligible signals, and was thus more challenging in the broader, more heterogeneous field of behavior genetics.

As an observer at Coast, I experienced firsthand some of the ways in which practitioners employed shorthands to signal shared understandings about animal modeling in informal settings. I described one such incident in my field notes that took place over lunch. After buying some food from the hospital cafeteria, I sat down to eat it with Dr. Lam, a postdoc in the Smith Lab who had set up a TV monitor in the lunchroom space and was manually scoring videotapes from one of his recent elevated plus maze experiments:

> I sat back and watched a few rounds while I was eating my salad. At one point in watching the video I commented, "That guy really likes the open arms" when the mouse that we were watching at the time seemed to be spending more time there than the others. There was silence, and then Dr. Lam said, "Don't say 'like.'" It took me a minute to figure out what he meant here, and then I realized that he was chastising me for anthropomorphizing the mouse. I asked him about this, and he said that he had been corrected before by other people for saying things like what I had just said. He said that you should never say things such as "the mouse likes the open arms" or "the mouse is less anxious," you should say things like "the mouse spends a higher percentage of time in the open arms" or "the mouse shows less anxiety-like behavior."

Dr. Lam's correction of my improper description of the test shows how researchers at Coast developed shared understandings of the epistemic scaffolds of particular animal models, and were trained to talk about behavioral testing in ways that conveyed these understandings. The short admonition not to say "like" communicated several important messages about the accepted local views of the epistemic capacities of the elevated plus maze. Avoiding talking about what the mouse "liked" allowed researchers to sidestep debates about whether they could claim to know anything about the internal mental state of the animal. Dr. Lam's insistence that we restrict our descriptions to observable behavior reflected these broader contestations about the scope of claims that animal researchers could make. And consistently calling the mouse's behavior "anxiety-like" rather than "anxious"—even over lunch—was a way of teaching me that he and other researchers at Coast believed that claiming the elevated plus maze could be used to measure "anxiety" was too risky, and that the epistemic scaffolding of this test needed to be lowered to a safer, more defensible height. The precise nature of the disjuncture between mouse and human was left unspecified by this language, but the cautionary message was clear.

Marking out particular statements as "anthropomorphic" was another way that researchers at Coast signaled shared understandings about behavioral testing and enforced community norms. As my field note excerpt above suggests, the ways that I talked about the mice when I first entered the laboratory generated frequent reproaches about my tendency to anthropomorphize. Identifying what exactly made a statement anthropomorphic (or zoomorphic), however, was initially difficult for me to discern. The researchers themselves moved fluidly between mouse and human experience as a matter of course in conversations in the laboratory. When describing the movements that mice made in mazes, they acted out those behaviors using their own bodies, peeking their heads around imaginary corners and elongating their necks and backs to create a resemblance to what they called the "stretched attend posture." They used linguistic formulations that blended mouse and human experience, talking about mice that "don't drink enough to blow over" or describing a particular piece of testing equipment as an Olympic balance beam. Researchers even invited me to take on the role of the mouse in an experiment as a way of explaining how particular tests worked. In one instance, I observed Ian, a graduate student midway through the program who was doing research on hormones and depression, as he was training mice to press a lever that would give them sugar water. He explained the procedure to me by inviting me to put myself in the position of a mouse in the experimental chamber, and he held out his clipboard as an imaginary lever for me to tap as he walked me through the experimental design.[5] It was all the more confusing for me, then, when simply saying that the mouse "liked" something generated such strong reactions.

Eileen Crist (1999) has argued that such slippery attributions of anthropomorphism are common. She argues that there is no hard-and-fast definition of what counts as anthropomorphism; rather, "its meaning is tied almost strictly to its aspersive connotations" (7). It is the "figurative, erroneous, or naive attribution of human experiences to animals" (7) that attracts criticism, she writes. Daston and Mitman (2005) likewise direct attention to the uses of anthropomorphism, showing how diverse human communities use anthropomorphic thinking to reason through scientific problems or enroll allies in conservation efforts. Researchers at Coast similarly used anthropomorphism to perform boundary work, marking out particular links between humans and mice as appropriate or inappropriate. Consider, for example, this exchange with Dr. Jackson, where she explained the problems of testing hungover mice in the elevated plus maze to me in greater depth:

SJ: When you're looking at, say, withdrawal-induced anxiety, then you have a problem. Because during withdrawal—imagine you've got a hangover, what are you going to do?

NCN: I'd be sitting the middle.

SJ: You'd be sitting, exactly.

NCN: I would be hanging out and waiting until my time was up and I could go back to my cage!

[laughter]

SJ: Exactly! And so mice that are withdrawing from alcohol don't move very much. So how do you interpret their behavior on that maze? You can't. So you then don't know if it's behavioral, they're not motivated, you don't know if it's malaise, they just don't feel well. Like every time they lift their head the room goes like that, you know? [acting out a spinning motion]. There's no way to interpret those data.

This might appear to be a clear instance of anthropomorphizing. But Dr. Jackson not only invited me to place myself in the position of the mouse in the maze as a way of reasoning through the problem, she even joined me in this activity. She also embodied the withdrawing mouse, physically acting out a mouse whose head was spinning from a rather nasty hangover.

What made this activity acceptable was that I appeared to Dr. Jackson to be imagining what it would be like to be in the maze *as a mouse*, not as a miniature human. This thought exercise was entirely compatible with the ethological argument for the validity of the elevated plus maze, which emphasized the importance of creating tests that took the species specificities and the natural history of laboratory animals into account. The difference between thinking as a mouse versus thinking as a human quickly became clear in the next few exchanges, as I began to describe what I might be seeing on the floor or an adjacent wall. Dr. Jackson cut me off and reminded me that mice have relatively poor vision and rely very little on their sense of sight, and that I needed to "think carefully and not anthropomorphize" in imagining what the mouse was experiencing. Speaking from this mouselike perspective required both a good deal of species knowledge and a good deal of practice, and engaging in these thought exercises helped researchers train new members of the field and identify shared positions with other researchers. Practitioners who failed to speak in these carefully circumscribed ways revealed themselves to be practitioners who did not understand or value the "ethological relevance" of the elevated plus maze test, and might therefore also hold risky positions on its knowledge production capacities.

Graduate students were frequently subjected to these kinds of reproofs because they were new to the field and actively being trained in its conventions. Near the beginning of my time at Coast, I attended a poster session that the department was holding to showcase its research to prospective graduate students. Current students from the department displayed posters describing their research projects, and the faculty and prospective students circulated through the hallway asking questions. I talked at length with Ian. He described his work using the "forced swim" test, a protocol where researchers place a mouse in a large cylinder that is partially filled with water and measure how much time it spends floating versus how much time it spends actively trying to escape. As Ian spoke, I was fascinated by how consistently he described the test as measuring "depression-related behavior," and by how easily such a cumbersome construction seemed to roll off his tongue. I asked him about it, and he said that these kinds of linguistic constructions were terms of art for the field and using them just came to him naturally. But, he continued, he actually believed in the test "more than he [let] on." While many animal behavior geneticists subscribed to the view that the forced swim test was a good predictor of antidepressant drug efficacy but did not have very strong "face validity," Ian did not believe that this was the case. To him, the kind of helpless behavior demonstrated by mice that just floated passively in the water rather than even trying to escape looked remarkably similar to the type of "learned helplessness" seen in people with depression. The more he used the test, the more convinced he was of the similarity between mouse and human behavior. "At this point, when I see that," he said, tapping on a picture on his poster of a mouse floating in the water, "I can't think anything but depression."

Despite his consistent use of the cautious language favored at Coast, it appeared that Ian had not fully absorbed the local views on what kinds of knowledge the forced swim test could generate. It had not gone unnoticed by more senior members of the department that Ian had become fluent in the linguistic conventions of the community without taking up the epistemic stances linked to these conventions. Dr. Smith told me later that he worried that Ian was anthropomorphizing the mouse behavior he observed in his forced swim test experiments, and had become practically "messianic" in his enthusiasm for this test. In a culture of cautious claiming, such zeal was not admirable. The fact that Ian seemed to be operating outside of the local norms of good scientific practice in a way that was potentially damaging to his reputation was not lost on him either. "I think there are classmates and other professors that don't know me so well that think that I'm just like in love with the forced swim test," he reflected to me several weeks later. "You know, cannot put it down, want to only talk about it, want to only do forced swim tests, and that's not the case." He agreed that forced swim test experiments alone were not strong enough evidence to make the case about depression that he wanted to make, and he told me about his plans for the other experiments he would conduct to provide additional support for his hypotheses.

SIGNALING SHARED UNDERSTANDINGS IN LARGER COMMUNITIES

Many of the techniques that researchers at Coast used to instill and assess shared understandings depended on face-to-face interaction. But the contemporary mouse community is large and diffuse, so much so that it is difficult to refer to it as a "community" in the same sense as the *Arabidopsis* or *C. elegans* research communities (Ankeny and Leonelli 2011). Scholars studying the history of model organism communities have argued that as these communities grew larger, it became more difficult to maintain shared norms because the interpersonal relationships used to enforce those norms began to weaken (Kohler 1994). Signaling shared understandings of the knowledge production capacities of animal models within the larger animal behavior genetics community and the even larger mouse research community was similarly a more challenging task. In the absence of the subtle linguistic conventions that certified one's ethological knowledge and signaled cautious epistemic stands at Coast, practitioners had to adapt their strategies for delineating insiders and outsiders to their community of practice.

The elevated plus maze is a good example of a test that Coast researchers felt had been "taken away" from the field (see chapter 2), and was being used for work it was never designed to support. To use Joan Fujimura's (1987) terms, the elevated plus maze offered a convenient "theory-methods package" for creating "doable" research projects on anxiety, which facilitated its uptake in fields well beyond animal behavior. As Lister (1987) noted when he first introduced the test, it has many advantages over other anxiety models. Not only does the test measure anxiety without using negative stressors (such as shocks or bright lights), it also requires no training period and only needs to be performed once to get results. It is almost impossible not to produce data using the test, because (barring a mouse who decides to jump off the maze) the mouse will always spend some percentage of time in the open arms and some in the closed arms. A researcher can easily purchase a commercially produced maze and a video camera for a few thousand dollars, and the availability of standard test protocols, commercially produced mazes, software packages for data analysis, all facilitate its movement and uptake. As a *Nature Protocols* article succinctly puts it, the test is "easy to use, can be fully automated, and valid results can be obtained in a short, five minute testing period" (Walf and Frye 2007, 327).

In such a context, the need to signal shared understandings about the epistemic capacities of the elevated plus maze was both more pronounced and more challenging. How could Coast researchers know whether an experimental report from an unfamiliar researcher was conducted and interpreted well, and therefore trustworthy? To address this problem, they adapted strategies developed in face-to-face interactions in the laboratory to look for signs of the experimenter's stance on behavioral testing in the thin description of testing methodology available in published papers. Ian explained one of his techniques:

> One of the things that I feel is a good indication is who are you citing. So there are people who will cite—for example, in the forced swim test, it doesn't make much sense to—[grabbing a paper from his desk and pointing] here, they're citing Porsolt et al. 1997a, and I did too in my last forced swim test article, but it doesn't make sense to use that as the reference for how you actually set up the experiment, because nobody does it like Porsolt. . . . I mean you might do this cursory nod to the original article, and that's fine, I do that, but I think that anybody who's actually basing their experimental preparation on his is probably a little bit clueless.

Graduate student Hannah explained that when she perused the literature, she looked for specific phrases in the articles that would indicate knowledge of mouse behavior and an awareness of common methodological issues for reassurance that she could trust the data. "If they say 'sound attenuated chamber,'" she told me, "that gives you at least some hope that it's not in the middle of an open lab with people working, you know, running gels right next to them." If she did not see these kinds of phrases, she tended not to trust the results. Hannah reasoned that if the author did not know enough about behavioral research to employ such signaling language in her methods section, then the author probably also did not know enough about mouse behavior to know how to properly set up behavioral experiments.

That researchers at Coast were typically looking for signs of caution in papers from unfamiliar practitioners suggests that new users may have a tendency to build up the epistemic scaffolds of models as they take them up, often to heights that the original developers of these tests find unsupportable. As the elevated plus maze has become more widely used, researchers at Coast worried that it has also become more widely regarded as simply a "test for anxiety," suggesting that the circulation of tools and techniques goes hand in hand with the inflation of their epistemic capacities. They were concerned that "inexperienced" practitioners who adopted the elevated plus maze would "pollute the literature" with data from poorly controlled test runs, or make overly broad conclusions about "anxiety genes" on the basis of a few experiments. These inappropriate uses of the test, they feared, would ultimately undermine the credibility of animal behavior genetics. In her textbook on mouse models, Crawley also explicitly linked the kind of cautious language I have described in this chapter to the maintenance of the field's reputation. She wrote: "Tread softly when approaching a mouse model of a human psychiatric disease. Investigators have no insight into whether a mouse feels "anxious" or "depressed." . . . Members of our laboratory are taught from day one to use cautious terminology such as "anxiety-like," "depression-related" and "relevant to schizophrenia." The credibility of our field depends on avoiding the impression that it is possible to create a comprehensive mouse model of a human mental illness" (Crawley 2007, 261–62).

By reinforcing epistemic scaffolds with new evidence or lowering them to a safer height, animal behavior geneticists aimed to produce scaffolds that could weather criticisms from within the scientific community, and perhaps also challenges from distantly related knowledge communities, such as other scientific fields, funders, or the public. Negotiations around the epistemic scaffolds of collectively used animal models were therefore not only a site where practitioners worked out how to use these scientific tools to generate molecular knowledge about behavior, but also where the scientific enterprise of animal behavior genetics was framed for broader audiences.

GENERATING FACTS WITH PROVISIONALLY STABILIZED ANIMAL MODELS

All of the activity I have described so far concerns the capacities of animal models rather than the production of genetic findings, but these methodological discussions also have implications for the facts that are eventually produced using those models. That this work was taking place both before *and* alongside the production of scientific findings was especially consequential for how researchers established facts in animal behavior genetics. Practitioners at Coast did use models such as the elevated plus maze in their research even though they thought of the tools as only provisionally stabilized, but this provisional status impacted their perceptions of the stability of the findings generated using those models.

STS analysts have typically described scientific techniques and findings as existing on a continuum of stability, from those are highly contested and provisional to those that are so well established that it is nearly impossible to challenge them. In this characterization, problems, contingencies, and moments of interpretive flexibility are erased as scientific findings and techniques solidify, acquiring "an air of inevitability" as they do (Sismondo 2009, 120). Latour and Woolgar (1979) and Latour (1987) argue that the status of a particular statement can be inferred from the degree of modality scientists attach to it. As scientists turn artifacts into facts, they gradually eliminate the linguistic markers of uncertainty surrounding particular statements. Rheinberger (1997) likewise characterizes facts and techniques along a continuum of epistemic interest. He argues that as particular phenomena become more predictable, what researchers once viewed as "epistemic things" begin to transform into stabilized technical objects. These established facts and techniques form the foundation from which new findings can be produced—today's provisional findings become tomorrow's scientific tools, creating the spaces in which new epistemic things can be deployed and manipulated (Rheinberger 1997, 80).

These analyses differ in their views of the fixity of these stabilizations. Latour and Woolgar (1979, 77) argue that once statements have solidified into taken-for-granted knowledge, researchers who continue to question those statements are "regarded as socially inept." Rheinberger (1997), in contrast, argues that an object's status as a technical thing is not one that is fixed for all time, and that established tools can transform into epistemic things when they are introduced into new contexts. Within the boundaries of particular communities or experimental systems, however, these theories both suggest (in broad strokes) that facts and techniques can be placed along a continuum of stability.

In the case that I have been describing here, however, the widespread use of the elevated plus maze for setting up experiments did not mean that it was devoid of epistemic interest for animal behavior geneticists. Rather, it seemed to occupy both ends of the continuum simultaneously, even within one community of practice: it was in use as a routine tool for investigating other research questions even while its own contours and responses to manipulation were still being investigated. While some researchers were using the elevated plus maze as a platform for researching the genetics of anxiety, other researchers were working away on the scaffold supporting that platform, adding in new links and testing the strength of existing ones.

Some in the animal behavior genetics community undoubtedly saw Coast researchers' ongoing inquiries into the methodological foundations of behavioral animal models as bad scientific manners, but Coast researchers saw it as responsible science. When I asked Emily about her colleagues' reaction to a recent paper she had coauthored on problems with the measurements of anxiety-like behavior used in several popular tests, she told me that the paper had attracted a lot of attention at conferences, particularly from senior scholars who had relied heavily on those tests in their research. I commented that I was not surprised, because her research could be seen as casting doubt on many years of established results. "Better to undo twenty years of science rather than fifty," she replied with a shrug.

These dynamics cannot be fully explained in terms of adversarial maneuvers. It was not advocates of rival animal models who were voicing concerns about particular tests; in many cases, it was the model developers themselves who were pointing out problems with their own tests and attempting to reinsert modalities into the scientific conversation. There was also no clear division between those who were using particular animal models as tools and those who were working on their epistemic scaffolds. At Coast, researchers actively worked to keep visible the limitations of animal models that they used in their own scientific work. Even after her research on measurement problems in anxiety tests, Emily still used the elevated plus maze as a tool in other aspects of her work—she just made sure, she told me, to be clear about the limitations of elevated plus maze data when she reported on her findings. Other animal behavior geneticists followed Emily's logic: they also told me that they regarded particular animal models as only provisionally stabilized and in need of further research, but in the absence of other more suitable tools, they often used them anyway.

This extrafactual work made it more difficult for animal behavior geneticists to see themselves as engaged in a project of creating enduring, translocal, transtemporal findings. The expectations for knowledge production that I described in chapter 1—that experiments with animal models would produce partial but useful findings that might aid clinical researchers in their search for treatments—were not just a product of the stories told by researchers at

Coast, or of their assumptions that mouse and human behaviors were complex. The abundance of methodological research and debate in the animal behavior genetics community also contributed to researchers' assumptions that their work would produce partial clues but not enduring explanations. How could practitioners make a strong claim about a specific gene's impact on anxiety when the tool being used to measure anxiety was itself subject to ongoing investigation and debate? Assumptions about complexity therefore impacted expectations about fact production both directly and indirectly—complexity narratives contained within them a set of expectations about what kind of knowledge animal behavior genetics research was likely to produce, and the methodological discussions that flourished under expectations of complexity further reinforced those epistemic expectations.

To address these perceived instabilities in their genetic findings, practitioners used triangulation between multiple tests as a way of compensating for the weaknesses of each individual test (a technique that resembled the way that practitioners combined multiple epistemic arguments to build more resilient scaffolds). Dr. Smith reflected on how triangulation became more common in the late 1990s as debates between molecular biologists and behaviorists about the interpretation of knockout studies intensified:

> If you looked at the beginning of the mouse knockout revolution, the early '90s, first half of the '90s, you'd see paper after paper after paper published in *Science* and all the hot journals that said we manipulated this gene, and then we put the animals in the elevated plus maze, and they showed decreased entries into the open arms, and therefore they were made anxious by this, and therefore this gene causes anxiety. You don't see that anywhere near as much anymore. Now what you see is the same experiment, but . . . you usually see that they've also tested them in the light/dark test, which is another emergence-based anxiety test. And you'll see that they've probably tested them in the open field, which is the original anxiety test that Calvin Hall developed in 1934. And if they get consistent answers in all three of those, then they'll say these mice are clearly more anxious. So, that's an improvement.

Presenting data from multiple tests was a way for those using models that were still under active discussion to increase the stability of their genetic claims by guarding against charges that their results could be attributed to the quirks or instabilities of a particular model.

Researchers used replication to do similar work. Dr. Smith told me he would only publish an experimental result after replicating it in his own laboratory three times, and he would only believe a result after someone else had replicated it. Practitioners seemed unusually enthusiastic about replication—far from being a boring, routine scientific activity, conducting replications seemed to behavior geneticists be valuable, interesting work. At one scientific conference, I was discussing a paper presentation with my hotel roommate Karen, a graduate student in a behavior genetics program in the Midwest. She agreed that the group's findings were interesting, "but it's only going to get *really* interesting," she told me, "if someone can replicate it." Karen continued on to tell me about how a senior scholar had stopped by that afternoon to see her poster, and this scholar told Karen that she had used many of the same measures in study that she did recently. The senior scholar advised Karen to frame the paper as a replication of her study when submitting it for publication, because it would increase the value of the publication. I was surprised that framing it as a replication would make the paper seem more important, rather than reducing the novelty of the finding. Karen shook her head. "This is behavior genetics," she told me. "One study means nothing." Others at the conference seemed to share her opinion. The next day, I saw a panel where several practitioners who were doing very similar work on nicotine addiction described their studies as "excitingly convergent." The final speaker in this panel concluded his presentation with a slide titled "Next Steps: More Replication." These attitudes toward replication resemble Harry Collins's (1985, 135) description of replication in a "poorly understood area," where "scientists just do not know enough to be able to guarantee that an experiment which looks just the same as another is the same in essence." It seems a stretch to say that the kinds of experiments I have been describing here constitute a poorly understood area, and yet Collins's description seems surprisingly apt. After thirty years of work with the elevated plus maze, researchers knew quite a lot about the test's behavior under different conditions, and yet still not enough that similar-looking experiments were uninteresting.

The simultaneous treatment of animal models for behavioral disorders as both established and provisional, research tools and research projects, changed both the scope of debates about findings produced using those models and the character of their resolution. A statement about a particular gene's involvement with anxiety may have been more convincing when it was supported by data from three models rather than one, and when multiple practitioners had arrived at the same conclusion. But the perceived instabilities in and ongoing work on the models themselves left the door open to future reexaminations of even these triangulated, replicated findings. With a larger part of the sociotechnical network supporting particular findings visible to researchers and in flux, the terms of the debate were not necessarily limited to the kinds of specific moves in the evidentiary chain of a particular finding that Pinch (1985) and Latour (1987) describe. There was much more available to be questioned. Practitioners were continually readjusting the network of tools and findings supporting their fact production, more deeply altering the profile of the terrain on which they were working.

CONCLUSION

The long-term stability of research programs depends on the presence of widely accepted facts, theories, and tools that practitioners can deploy in their research. My aim in this chapter was to show how a series of small disputes about the elevated plus maze could be thought of as part of a larger process of negotiating the epistemic foundations of the animal behavior genetics enterprise. What was at issue in these disputes was not just whether a particular researcher could claim to have found a gene for anxiety (or merely a gene that influences one aspect of anxiety-like behavior); it was broader questions about how to do the work of animal modeling. Should researchers use the elevated plus maze to do studies of genetic knockouts or withdrawing mice, or should they restrict themselves to doing preliminary screens of potential anxiolytic drugs? If they used the maze for conducting knockout studies, should they be able to publish claims on the basis of these experiments alone, or should they be required to conduct more than one anxiety test? These are questions about the routine practices of experimentation and claims making in the field, not the modality of individual statements. Even in cases that seem at first glance to concern the use of linguistic markers of uncertainty, I have argued that these linguistic markers matter to researchers because they act as indicators for more general positions on knowledge production.

Reframing these disputes as part of a collective process of building a research program helps to make sense of actions that seem illogical when seen through other analytical lenses. In a view that portrays laboratory life as a competitive struggle to establish specific facts, it is difficult to understand why researchers at Coast would spend precious time and resources conducting experiments that question an animal model that they use in their own research. But when considering the status of the field as a whole and its long-term stability, these actions become more comprehensible. Negotiations about the epistemic capacities of animal models were techniques for responding to matters of collective concern, such as how to establish sustainable scientific practices or how to retain credibility in the eyes of funding bodies or activist organizations who questioned scientists' claims about their knowledge making tools. These field-level concerns were especially important in animal behavior genetics, where practitioners were particularly concerned about the public perceptions of their field and the potential harms of making expansive knowledge claims. Negotiations about the capacities of animal models were also a site where understandings of behavioral disorders and of the humans who live with them were framed through the process of research. As I will explore more in chapter 5, in the process of developing a model and its epistemic foundations, and through a series of often quite pragmatic inclusions and exclusions, researchers created different spaces of possibility for representing human behavior.

My aim here was not to make my own evaluation of the merits of the elevated plus maze as a model for studying human anxiety, but to show systematically how practitioners themselves made and managed associations between mouse experiments and human anxiety. Although it could be advantageous for researchers to make strong claims about the capacities of mouse experiments as knowledge production tools, researchers did not want to completely collapse the distance between animals and humans. By excluding aspects of human disorders as inappropriate for animal researchers to study, or using language that suggested uncertainties in the relationship between animal models and human disorders, animal behavior geneticists navigated contentious disputes both inside and outside of their scientific community. And by making, managing, and breaking specific links in the epistemic scaffolds of their tools, animal behavior geneticists carefully negotiated the strength of the claims that they made about animal models and human disorders.

CHAPTER 4

Epistemic By-Products: Learning about Environments while Studying Genetics

About three weeks into my stay at Coast, Kimberly, one of the Smith Laboratory technicians, invited me to be the record keeper for one of the studies that she was running. The experiment was a "loss of righting reflex" study. Researchers administered doses of alcohol to their mice that were large enough to nearly immobilize them, and then placed them on their backs in a V-shaped trough. Then they measured how long it took the mice to recover enough that they could flip over onto their paws and attempt to run away. The study was a large one and called for all hands on deck—all four of the Smith Laboratory technicians were pressed into service. They arranged the tables in a ring around the outside of the procedure room, and set up supplies and cages of mice at each technician's station. I sat in the middle of the room where I calculated the correct dose of alcohol for each mouse based on its body weight, and called out the figures to the technicians as they prepared the injections. Soon, over a hundred very inebriated mice lay upside down in the troughs surrounding us, their soft underbellies rapidly rising and falling.

While we waited for the first mouse to make its escape attempt, the conversation turned to complaints about the newly adopted colony management software. The program was a publicly available version of the Jackson Laboratory's own system. The software assigned a unique number to each mouse at birth so it could be tracked individually as it moved through the facility, rather than tracking groups of mice by cage number as their previous software system did. Tracking individual mice, however, created six times as much data entry work. Kimberly complained that she was currently three generations behind on entering the data for one of her colonies. Ironically, mice that they purchased directly from the Jackson Laboratory came with little identifying information. Other than the strain, the Jackson Laboratory typically provided only the room number or sometimes just the building number identifying where the mice were born.

I asked whether this level of detail really mattered, because the mice were all theoretically exactly the same. "But they're not," James, another one of the Smith Laboratory technicians, corrected me. He explained that of course the mice were not exactly the same—or else there would be no point in testing a hundred mice to identify the loss of righting reflex time for that line, he said, gesturing to the mice that surrounded us. These mice were genetically similar because they had all been bred for their tendency to drink heavily in particular experimental situations. But no one expected the mice that surrounded us to roll off their backs in synchrony, he continued—and imagine the chaos that would ensue in the procedure room if they did! The remaining genetic differences between them could be responsible for differences in their physiology, as could differences their early life experiences or recent experiences in the housing room. To drive home the point further, James described a different mouse strain whose fur could range from a bright yellow to a dark brown depending on its early environment, even though all of the mice were genetically identical.[1]

The animal behavior genetics laboratory is by definition a site for producing knowledge about the genetics of behavior, but it can also be a site for producing knowledge about individual differences and nongenetic components of behavior. Although the experimental work at Coast aimed to make visible particular kinds of genetic difference, doing so required sustained attention to other kinds of difference. The goal of the study I participated in was to calculate the average loss of righting reflex time for the Smith Laboratory's selectively bred mouse line, which they then planned to compare to the average for their control mouse line. They wanted to see whether the genes responsible for the difference in drinking between the selected mice and the controls also generated differences in the way those mice processed alcohol. But in order to see whether the genetics of heavy drinking was related to alcohol metabolism, Kimberly and her fellow researchers had to first make visible the substantial amount of variation between individual animals. In other words, they needed to generate knowledge about differences that they did not intend to publish on in order to generate knowledge about the ones they did.

What happens to this knowledge about differences between animals, or the impact of the environment on the mouse? Why would James deliberately draw my attention to all of this variation, which seems to call into question the stability of behavior? This chapter examines how researchers managed the "epistemic by-products" that they accumulated while carrying out their research. In many respects, this management work resembles the scaffold work I described in chapter 3. It is extrafactual work that takes place before and alongside the production of scientific facts, and that has consequences for the stability of those facts and of long-term research programs. The way that researchers at Coast managed the knowledge they gained about the laboratory environment was both a reflection of their assumptions about complexity and something that perpetuated those assumptions—this work helped them create controlled environments they desired to study the genetics of behavior, but it also reminded them of the radical limitations of such an endeavor. Some researchers also managed nongenetic knowledge quite self-consciously with the aim of shaping collective knowledge production practices. By drawing these by-products into new infrastructures, deploying them to explain and preserve difference in findings, or designing experiments to transform them into more acceptably scientific forms, researchers made both general and specific interventions into their fellow researchers' experimental practices and interpretations of their findings. In this way, the management of epistemic by-products allowed researchers to fine-tune the claims that they and others made about mouse experiments and human disorders.

GENERATING KNOWLEDGE ABOUT THE LABORATORY ENVIRONMENT

As I argued in chapter 1, life in the laboratory at Coast was characterized by an intense degree of concern with the sensory aspects of the laboratory environment, such as the objects, sounds, and smells that the mice encountered in their home cages or in the testing rooms. Variation in the laboratory environment caused concern for researchers because they saw a controlled environment as essential for making visible the relatively small genetic effects they expected to find. While it was the genetic inputs into behavior that they expressly sought to know more about, they believed that they also needed to understand and manage other inputs into behavior to accurately interpret their findings.

Managing the laboratory environment at Coast was quite time consuming, both because of the scope of what counted as "environmental" for researchers and the kinds of variations they believed might be significant. The environment was a residual category for them, one that collected up a variety of entities and actions. When I first arrived at Coast, for example, I was surprised that researchers referred to themselves as "environmental factors." In what way, I wondered, were humans "environmental"? The answer, simply put, was that the experimenters were external to the mice, and so they constituted part of the environment that the mice experienced. The laboratory environment, then, could include almost any factor that was external to an individual animal, and researchers at Coast seemed willing to entertain the idea that almost any variation in these factors could impact behavior.

Controlling the laboratory environment occupied a proportionally large amount of their attention because Coast researchers assumed that other aspects of their experimental systems were less consequential or already well controlled. In particular, they regarded many aspects of the mouse genome to be well controlled, thanks to the efforts of organizations such as the Jackson Laboratory that monitored and maintained the "genetic purity" of inbred strains. When they talked about genetic control, researchers might add the caveat that even genetically identical mice could develop spontaneous mutations, but they saw these mutation events as known entities that occurred at a calculable and predictable rate, and were therefore easy to anticipate and manage in the research setting.[2] Strong social and technological commitments to the stability of the mouse genome in the broader research community reinforced this understanding. Questioning the genetic identicality of inbred strains of mice would have been costly, because it would have jeopardized many established researchers, institutions, and research results.

Consequently, in their everyday practice, researchers at Coast spent as much (if not more) time attending to the laboratory environment as they did to the genetics of their mice. And when something unexpected happened in an experiment, they almost always looked to the environment, rather than the genome, for the source of the anomaly. New sources of genetic variation were not unheard of. The discovery of genomic "copy number variants"—small duplications in the genome that could result in multiple copies of a gene—was a hot topic of discussion while I was doing my fieldwork at Coast, because some evidence suggested that these genomic variations could be responsible for variation in anxiety behavior (Williams et al. 2009). But such discoveries were rare. Indeed, one of the things that made genetic facts so valuable was their relative rarity compared with the other kinds of knowledge the laboratory generated. To researchers at Coast, it seemed much more likely that differences in the laboratory environment caused the many fluctuations and individual differences they observed in the behavior of their mice. In contrast to genomic variation, evidence of environmental variation seemed to be all around them, from the clanging carts that sporadically passed through the hallways to differences in the researchers' own demeanor from day to day.

Under these conditions—where researchers understood behavior to be complex, and saw the genetics of their animals as tightly controlled but the environment as highly variable—the laboratory became a place where researchers could learn about both genes and the environment. Working with genetically identical animals that nonetheless differed in noticeable ways reminded researchers that the environment could substantially modify behaviors that were under some degree of genetic control. And as researchers enacted controls and repaired breakdowns in the housing and testing rooms, they had ample opportunity to observe the impact of specific nongenetic factors on behavior. Because so much of their daily work revolved around experimental control, researchers arguably collected many more observations about the effects of environmental factors on behavior than they did about the effects of genes.

Much of the knowledge generated while managing the laboratory environment remained highly anecdotal. Technicians at Coast were especially quick to draw conclusions about the effects of their own handling techniques or other changes, and they tended to convey this information through stories about their personal experiences. The Martin Laboratory employed several technicians who had been with the lab for many years, and these technicians were quite vocal about how noises, techniques for handling the animals, or even the temperament of the experimenter could affect the behavior of the mice. Susan, who had been working for Dr. Martin for more than a decade, described the mice as highly responsive to her mood. When I shadowed her for a day, she stopped me in the hallway before we went into the first mouse room and told me that she always took a minute to calm herself before she went in. She could see the difference in how the mice behaved, she said, when she was distracted or in a hurry. She noticed the same effect when her husband helped her make rounds on weekends. He hated the smell of the mouse rooms, and he quickly got flustered as she read off the figures from the mouse cages for the record book. Susan said she could see how his discomfort impacted the mice. And so, we paused for a moment of silence and a few deep breaths before proceeding to check the levels in the alcohol bottles on the cages in the first mouse room (which Susan read off at such a rapid pace that I became overwhelmed as well, and I had to stop and ask her to back up several cages so I could make sure I had the correct figures).

Researchers at Coast also speculated about how the environment impacted mouse behavior, but unlike the technicians, they tended to be more circumspect in their conclusions. I heard countless admonitions from researchers about smell and noise in my travels through animal behavior genetics laboratories. However, rarely did researchers claim to see an immediate impact on the animals from these factors, and the kinds of assertions about mouse behavior that technicians voiced made the scientists uncomfortable.[3] Researchers' observations had more of a precautionary quality, where clear evidence of an effect on the animals was not needed to make recommendations about how to do science. Rather than making specific assertions about which factors might impact their mice, they talked about how their general knowledge of the mouse's sensory capacities made it seem plausible that a ringing cell phone could disrupt their experiment.

But not infrequently, situations arose that looked so much like experiments that even the researchers began to draw conclusions about specific environmental variables. One such situation came up in an interview with Ava, a graduate student in the Martin Laboratory. She told me about her experience working on a large project where the laboratory was developing lines of mice that differed in their response to a stimulant drug. The Martin Lab was interested in "sensitization" to the drug, a behavior that is the opposite of tolerance—the tendency for an animal to show a stronger response to a drug the more times they receive it. Studying sensitization required that researchers test all the mice several times to see how their response to the drug changed over time. The length of the study, Ava recalled, made it difficult to keep all of the factors constant that she might have otherwise wanted to control for: "This selection for sensitization, and it's a multiday study, and you know, I still had classes, and so there were a lot of times when it just wasn't feasible for me to do it. And so [a technician] would do it, and she always got way more robust sensitization than I did. But the question was always why. And then, I got dogs! And my response came up." Having a technician run the study for her on days that she had to go to class provided opportunities to observe how the same mice, under the same experimental conditions, behaved differently depending on who was running the experiment. It quickly became evident to her that there was an effect of the experimenter on the mice. The relevant difference between herself and the technician remained a mystery, though, until her results shifted around the time that she adopted a dog. Because the technician who helped her run the experiments also had a dog, Ava then concluded it was the smell of dog that affected the behavior of the mice (a conclusion no doubt informed by the importance placed on sensory aspects of the laboratory environment and ethological understandings of mouse behavior at Coast). She admitted to me that this explanation for the change her in data was not exactly a "scientific explanation," but it nonetheless informed her experimental practice. After this experience, Ava told me that she started recording "way more information than [she] ever used to," because she saw more clearly how environmental variables could alter her experiments. Ava also worked to preserve and circulate this knowledge, reconstituting her laboratory experiences for me so that I could also share in her realization about the impact of certain smells on mouse behavior.

Researchers worked in other ways to gather observations about the laboratory environment and transform them from personal anecdotes to collectively held knowledge. During the same set of sensitization experiments, a conversation at a weekly meeting in the Martin Laboratory inadvertently revealed another difference. The protocol that Dr. Martin originally developed required the experimenter to place the mice in temporary "holding cages" after she had weighed them but before they were tested.[4] While giving a progress report on her research, one student mentioned that she weighed each mouse individually and placed it directly in the testing chamber, thereby bypassing the holding cage step. Members of the laboratory and even other senior researchers in the department had conflicting opinions on which version of the testing protocol they preferred, so Dr. Martin had one of her students conduct an experiment on the question. After systematically comparing the results of zero, five, or ten minutes of holding cage time, the Martin Laboratory found that omitting the holding cage step resulted in a less robust sensitization effect. By formalizing the knowledge through experiment, it became easier to convince all of the members of the laboratory to adopt the same protocol, and made them more confident about passing on this knowledge to other laboratories. When other scientists called to inquire about their testing methods, the Martin Laboratory manager told me she always made sure to mention the importance of using holding cages.

EPISTEMIC BY-PRODUCTS IN LABORATORY WORK

The kinds of observations I have pointed to so far—that an experimenter's harried demeanor might agitate the mice, or that moving a mouse to a temporary cage might be crucial for making a sensitization experiment work—are things that science and technology studies (STS) scholars might typically classify as tacit or craft knowledge. The idea that laboratory work requires and generates many different kinds of facts, observations, and skills was one of the foundational insights of early laboratory ethnographies. In his pioneering study of the transversely excited atmospheric (TEA) laser, Collins (1974, 1985) drew on the concept of tacit knowledge to highlight the embodied technical skills that were crucial to building a functioning laser. He traced out several British laboratories' attempts to build their own version of a laser first built by a Canadian group and found that the key to their success was socialization in the art of seeing relevant similarities and differences between lasers. Michael Lynch (1988) has argued that embodied skills and tacit "know-how" are similarly crucial in the animal laboratory. Lynch described, for example, how researchers in the laboratory he studied needed to learn how to soothe or disorient their rats before giving them injections so the rats would not curl up on themselves and bite at the needle.

The analytical questions raised by the tacit/craft knowledge framework center on the attributes of this knowledge and the way that it travels (or does not travel) through knowledge communities. Is tacit knowledge different in kind than the knowledge that appears in scientific texts, defined by its inability to be successfully communicated through diagrams or written descriptions? Or is it simply that which scientists do not reveal in their writing? Collins (2001, 2010) has offered a detailed classification of different types of tacit knowledge, and distinguishes these types from "concealed" knowledge that could in principle be transmitted through writing, even if it is not in practice.[5] His formulation raises additional questions about the reasons why scientists might not communicate particular kinds of knowledge. Collins (1974) suggested that competition between research groups was one reason that researchers might refrain from describing important tips and tricks for getting experimental setups to work. Alternatively, scientists may not be aware of the extent to which they rely on tacit/craft knowledge in their own work.[6] Or they may recognize its instrumental value for accomplishing particular tasks, but view it too idiosyncratic or unproven to merit formal publication.[7] Researchers may also be trained to "launder out the tacit" from public accounts of their research (Delamont and Atkinson 2001, 102) to create the stylized narratives that are expected in articles, conference presentations, and other public venues.

These analytical questions and classificatory schemas make it more difficult to see how this knowledge can fulfill multiple functions and how its value varies with context. This chapter puts forward an alternative framework for thinking about the unpublished knowledge generated through laboratory work, one that does not rely on the assumption that it is different in kind from what appears in print. Instead, I define knowledge about handling, housing, and temperament as "epistemic by-products"—observations that researchers accumulate as part of the process of carrying out what they consider to be their main line of knowledge production work. Rather than defining epistemic by-products with respect to particular intrinsic features, I define them relationally, with respect to particular experimental aims and programs. This relational definition has the advantage of not fixing the value of particular observations—not only can what counts as a product or by-product vary substantially, but whether a by-product is considered waste or valuable depends on who is evaluating it, or the presence of technologies or markets that could transform it or facilitate its movement.

To illuminate these knowledge production dynamics, I rely on the metaphor of the laboratory as sawmill throughout this chapter. Just as the process of cutting and planing a piece of wood to create a board necessarily generates sawdust and wood chips, the process of constructing scientific facts inevitably produces other entities. For some factories, producing wood chips may be simply a side effect of making lumber, while for another it might be the central purpose of their operation. Under some circumstances, the production of sawdust might be the opposite of valuable—a sawmill might have to pay for its disposal. But with a shift in thinking or economic conditions, that same entity could potentially generate revenue. An enterprising sawmill owner might turn her wood chips into a product for landscaping, or changes in the price of lumber might bring other entrepreneurs to the sawmill in search of sawdust for making wood stove pellets or IKEA furniture. Sawmills may necessarily produce sawdust and wood chips, but how these are entered in the business ledger or circulate in markets is far from fixed.

Likewise, what might be considered epistemic waste in one context might be seen as desirable to know in other locations. Helen Longino (2013) describes the behavioral sciences in this way: as sets of overlapping research traditions that each "parse" the same "causal space" differently. What distinguishes these approaches, she argues, is that some hold constant what for others are the very things to be investigated. A molecular behavioral approach might involve standardizing the environment in the service of capturing genetic variation, while a social-environmental scientist would do the opposite. Hannah Landecker (2013) argues that the value of particular observations might vary even for the same researcher over time, where things that might once have seen as a nuisance can develop into valuable new lines of research. Landecker describes how one researcher's observation of fluctuating hormone levels in her control mice—something she initially saw as an impediment to her experimental work—led her into a new research project on the physiological activity of bisphenol-A (BPA). Recent studies of "value practices" in the life sciences (Dussauge, Helgesson, and Lee 2015) also emphasize the multiplicity and flexibility of the assessments that actors make about what is known and what is worth knowing. Researchers may value an observation for its profitability at the same time that they value its ability to enhance their scientific reputation or promote access to medical care. Rather than trying to separate out different kinds of economic, cultural, and social values, Dussauge and colleagues (2015, 3) argue that we should treat value as "multifaceted, shifting, and entangled." It is these properties I aim to emphasize by choosing the term *by-products* to describe what researchers acquire in the course of pursuing their research agendas.

The materiality of the sawmill metaphor also has the advantage of keeping the work required to attend to epistemic by-products firmly in view. Sawdust is not something that disappears from view if the sawmill owner does not value it. She cannot simply choose to pretend it does not exist—even if she has no interest in transforming it into a new product, it still accumulates in the machinery and on the workshop floor and requires effort to clean it up. Visual tropes, such as those of visibility/invisibility, can disguise this work, making it seem as though researchers can simply ignore what they do not value.[8] Landecker (2013), for example, nicely articulates the historical and material circumstances that were important to the emergence of a new research program on BPA, such as increasing public attention in the 1990s to so-called endocrine disruptors and the way that they intensely controlled space of the laboratory brought "background assumptions into view." She ultimately argues, though, these factors themselves could not bring about these transformations because researchers still needed to direct their attention toward potential environmental harms. In a historical moment where genetics draws so much focus, she concludes that "we know next-to-nothing—because we have thought next-to-nothing—about how [mice] have been fed and housed."

Scientific practitioners who believe their fellow scientists ignore the importance of the laboratory environment similarly describe the problem in terms of visibility or attention. Hanno Würbel, a Swiss ethologist who specializes in laboratory animal welfare, described to me the movements to "standardize out" the effects of the laboratory environment as follows:

> It's kind of—it's almost schizophrenic in a way, because standardization is based on the realization that the environment affects your results. Because you find that the environment affects your results, you try to standardize it

> so that it can't affect them anymore. By this of course you basically try to get rid of the environment. . . . It's like—it's as if people attempt to spirit the environment away.

Würbel told me that he thinks that most biomedical researchers treat animals as "test kits," just like pH strips or other laboratory equipment. In doing so, he believed that his colleagues (perhaps purposefully) ignored how animals change in response to their environment. He described his own experimental work as an attempt to make the environment visible again by showing how the behavior of laboratory animals depends on their surroundings.[9]

To control something, however, is not synonymous with making it disappear. Variation is not "spirited away" by supernatural forces; researchers and technicians reduce variation through painstaking effort. Studies of tacit knowledge make this clear, as do studies that attend to the work and resources involved in "unknowing" or the creation of ignorance,[10] and studies of how researchers attempt to manage the liveliness of animals in experimental settings.[11] As Gail Davies (2013, 148) puts it in her study of mutant mice, "The work to articulate experimental assemblages, to make matter speak, is equaled by the work done to keep matter silent." To say that "we know next to nothing" of how mice are fed and housed, then, is not quite right (depending on how one interprets the "we" in this sentence). Researchers need to know about these things in order to work with mice in the laboratory, although they need not see that knowledge as valuable in the way that Landecker's (2013) protagonist did. The question, then, is not whether researchers pay attention to these things but how they attend to them.

Recognizing the ubiquity and the flexibility of the knowledge that researchers acquire through their work is important for understanding how "tacit" or "craft" knowledge can be more than just a set of tips and tricks for getting experiments to work. Researchers at Coast took knowledge about nongenetic factors quite seriously because they viewed them as part of the complex system they were studying. In this view, observations about the impact of a dog's smell on mice counted as knowledge about behavior, just not the kind of knowledge that they believed would lead to new clinical interventions. But even researchers who did not view behavior as complex still needed to manage smells and sounds in their own laboratories, and their evaluations of these factors could not be made completely independently from those of other researchers. Like debates around epistemic scaffolds of animal models, the management of epistemic by-products was a collective project, a site where researchers negotiated the meaning of the various entities the laboratory produced.

THE VARIABLE VALUE OF THE LABORATORY'S PRODUCTS

Observations made in the laboratory can have a high degree of interpretive flexibility, allowing them to take multiple forms and serve multiple functions. For researchers at Coast, observations about the effect of a dog's smell or a holding cage on mouse behavior were valuable because they helped make their experiments work, but they were also useful for evaluating other researchers' models and findings, and enhancing their credibility in a community that valued experimental control. In certain circumstances, researchers decided that knowing about a particular environmental factor was valuable enough that they worked to transform that knowledge into a more recognizably scientific form. In some situations, what was once an epistemic by-product might rise all the way to a highly valued (and publishable) biomedical finding.

One of the stories that I heard at Coast demonstrates how the scientific value of particular observations could vary substantially for different researchers. Ava described a seminar presentation from a faculty member in the department who had been doing research on "novelty-seeking behavior" with mice. This professor had a reputation among the behaviorists for being an "anatomy guy," and even though he was a respected colleague, they did not consider him to be part of their research culture. The professor gave a presentation on correlations between tendencies in his mice to explore new objects in their cages and differences in their neural pathways and described the lab's future plans to see whether there was a genetic basis for these differences. However, his lab had been using C57 Black 6 mice for their experiments, one of the most popular inbred mouse strains, and he had seemingly not registered that this meant all of the brains he was looking at came from genetically identical animals. The differences he saw, then, could not be due to genetic variation. The behaviorists in the seminar room still found his data quite interesting, noting that this offered an opportunity to look for environmental factors or developmental mechanisms that could produce such differences in neural pathways. But their neuroanatomist colleague lost interest in following up on the observation once he realized that he could not relate those brain differences to genetic differences. Ava's purpose in telling me this story was to draw attention to what she saw as a dangerous lack of knowledge about genetics in the broader neuroscience community, but this story also illustrates how a single observation can have very different degrees and kinds of value—brain differences might be potentially valuable biomedical knowledge, worthless red herrings, or perhaps indicators of a poorly controlled laboratory environment.

The Mouse Phenome Project provides another example of how some researchers might see scientific value in observations that others regard as by-products. The project, which is housed at the Jackson Laboratory, is an effort to collect information on the baseline physical and behavioral characteristics of different inbred mouse strains (such as body weight, blood pressure, or activity levels) and on the protocols researchers use for breeding and testing those mice (such as housing conditions, diet, or test parameters). An interdisciplinary group of mouse researchers first discussed the concept of such a database at an informal meeting at the Jackson Laboratory in 1999, and they officially launched the project in 2004 (Bogue and Grubb 2004; Paigen and Eppig 2000).

By building a database to organize and connect different types of knowledge, the Mouse Phenome Project aims to turn observations collected in the service of producing other facts into something with new value. When I visited the Jackson Laboratory in 2008 to talk with Molly Bogue, the director of the project, she explained that researchers routinely collected data on their control animals and the environmental conditions in their laboratories to calculate the effect of particular experimental interventions. But there was no easy way to compare these baselines across publications to see potential patterns, because these observations were embedded in standalone academic papers (if they were published at all). She wanted to encourage researchers to contribute their baseline readings and protocol information to the database where they could be put to good use—to deposit their by-products in a kind of epistemic recycling bin, rather than simply using them once and then throwing them away. Aggregating and organizing measurements of baseline behavior, physiology, and the laboratory environment, she argued, could make those measurements useful for new purposes: researchers might be able to see how behavior varied with environmental change, select more appropriate strains for drug discovery or toxicology studies, or identify new correlations between genes and physiological biomarkers.

These examples demonstrate the analytical problems with making distinctions between different types of knowledge in the laboratory. In these cases, the distinctions between product and by-product, method and finding, technical knowledge and scientific knowledge were less a property of observations themselves and more a product of the specific contexts of action that the observations were embedded in. When looking at a specific researcher or moment in time the distinctions might seem fairly clear, but when considering the broader context of a research program the boundaries quickly become messy. What might be a by-product at one time could become part of a publication later on. Even in the same laboratory and at the same moment, these categories could be partially overlapping. The Smith Laboratory, for example, worried about noises and handling procedures because they believed it was important to control for the stress levels of their animals, and so knowledge of stress was instrumental to executing their laboratory work. But in a few projects, stress was something that they deliberately manipulated to generate data on how stress impacted drinking behavior. While stress research was something of a side project for the Smith Laboratory, stress and behavior was the main line of research for one of Dr. Smith's colleagues, and the National Institute on Alcohol Abuse and Alcoholism also considered stress research valuable enough to fund a multisited, decades-long initiative on stress and excessive alcohol consumption. Observations about stress and behavior were thus products and by-products, craft knowledge and scientific findings, not only within the same research community but even within the same laboratory.

Certainly, there were some stable patterns in what researchers valued and how they made distinctions between observations. Whether a finding might be publishable (and in particular, publishable in a high-profile venue) was one factor that they used to assess the many observations that they gathered in the course of their work. Another related criterion that researchers used was the potential relevance of a finding for human health. When I asked Bogue to tell me about the potential uses or outcomes of the Mouse Phenome Database, she described the project in a way that emphasized translational value:

> One of the major goals of a project like this is to end up with these huge data sets, and someone comes in, a statistical person or a computational biologist, they come in and they just take all the data. They don't care what it is, they don't care what the protocol is, they want it, and they just do a huge, global analysis of that data and crunch it to try to find things that make sense, and they find patterns that are interesting, or something about it that's interesting to follow up. That's when they are very interested in the protocol, so they can go back and say oh, this is clustering with this under these conditions but not those conditions. What's different? Oh, this is a high-fat diet, and this is a low-fat diet. Oh, wow! So this gene pathway is involved in [this disease], and when there's not a high-fat diet you don't see that.

Bogue's hypothetical finding makes it easy to imagine how information about mouse housing could be connected to phenotype information and translated into a new clinical intervention. The original paper outlining the concept of the Mouse Phenome Project used a similar example with obvious clinical relevance—a set of studies on the connection between salt and hypertension in rodent models (Paigen and Eppig 2000, 715).

While researchers at Coast generally saw genetic findings as easier to relate to human health than nongenetic ones, taking a longer historical view upsets the idea that this is a fixed property. There is no intrinsic reason why it would be more difficult to see findings about the laboratory environment as having relevance for human health than genomic ones. Edmund Ramsden (2011) has shown, for example, that John B. Calhoun's experiments on rodent housing environments in the late 1950s were widely referenced in psychology, urban sociology, and the design professions as a model for understanding the impact of urban spaces on human behavior. Rebecca Lemov (2005) has similarly demonstrated how experiments on animal behavior and environmental stimuli captured scientific and popular imaginations in the Cold War period, serving as a substrate for visions of human behavioral engineering. Findings that behavioral researchers today might view as useful only insofar as those findings help them design genetic experiments might have been seen by researchers at midcentury as the keys to reforming human society.[12]

THE COLLECTIVE MANAGEMENT OF EPISTEMIC BY-PRODUCTS

In making distinctions between products and by-products, the perspectives of individual researchers matter, but these positions alone do not determine the fate of particular observations. The management of the laboratory's products is in some important respects a collective process that depends on shared infrastructures. A sawmill owner may be interested in finding an alternative use for her sawdust, but without systems for transporting, distributing, and marketing those by-products, she is unlikely to succeed. With the right infrastructures, however, the by-products of farms, factories, or slaughterhouses may become as valuable as the products themselves.[13] Likewise, a researcher's ability to treat some observations as irrelevant while treating others as scientific facts depends on multiple evaluations that are brought into contact through shared scientific infrastructures. While researchers manage their epistemic by-products in accordance with the assumptions and practices of their local communities, they are also subject to larger "market forces" that shape the meaning and value of those by-products, and by extension, of their scientific products as well.

Infrastructures such as the Mouse Phenome Project were one site where collective negotiations about the value of particular observations took place. The project brought together practitioners from a wide variety of scientific fields: behavior genetics, toxicology, cancer research, and commercial drug development, to name a few. Developing standards of reporting across these diverse communities unsurprisingly revealed substantially different understandings about what information was useful to collect. The project developers settled on what they thought was a "fairly minimal" set of reporting requirements about the laboratory environment, such as the type of bedding and the temperature of the colony rooms. Bogue recalled that she thought these variables were "no brainers" to include in the database, because in her view it was widely known that these factors varied between laboratories and could affect experimental outcomes. But when researchers began submitting protocols, Bogue told me that not all of the researchers submitted information on these variables. The database curators then began to fill in the gaps by recontacting researchers: "If we don't have the information that we think we need, we keep going back to the investigator. Which, you know, that can have problems too. The ones that don't really appreciate this detail or the need for the detail are kind of like, 'What is these people's problem? [laughter] Why do they keep bugging us, you know? Is it important that vitamins have been supplemented in our water? Is it really that important?' Well, yeah, we need to record that." Bogue explained that they tried to work with researchers who "cared about the details" while developing their submission guidelines in the early days of the project. They feared that researchers who did not "appreciate" these details would simply throw up their hands in frustration and walk away when the database curators kept e-mailing them.

Another way to view the response of these researchers, however, is as a form of resistance to having vitamins treated as a factor that could meaningfully impact the outcome of their experiments. It is not simply the additional work involved in collecting such measurements that might generate conflict, but the implication that those who have not been recording and reporting on this information have been neglecting to account for factors that are "known" to impact mouse physiology and behavior. The Mouse Phenome Project's efforts could be seen as diminishing the authority of the laboratory as "truth spot" (Gieryn 2006) by altering the value of knowledge about environmental difference. Kohler (2002) has argued that the laboratory is a place that gains its epistemic authority by virtue of its "placelessness." "It is simplified and standardized," he writes, "stripped of all context and environmental variation. . . . It is this odd spatial quality that gives knowledge produced in labs its credibility" (191). In practice, of course, laboratories vary in myriad ways, but researchers can maintain a sense of placelessness through their management of these differences—for example, by "disposing" of differences through arguments that they are not significant. The Mouse Phenome Project's alternative management strategy, which treats differences between laboratories as having potential epistemic importance, puts them into conflict with researchers whose strategies for creating credible facts depend on treating that same information as without epistemic value.

The flexible value of the laboratory's products creates other such points of contact where conflicts over their management can emerge. While knowledge about the behavioral impacts of dog smells, temporary cages, or different experimenters was a by-product of genetic inquiry for researchers at Coast, it was the main line of work for other scientists. Hanno Würbel, the Swiss ethologist interested in animal welfare, has made a career out of studying such differences.[14] His research examines the impact of environmental variables such as cage size on the mouse's behavior and brain development. By manipulating the environmental parameters that other practitioners seek to standardize, Würbel's work similarly poses problems for the credibility of other practitioners' scientific findings. In a series of studies conducted in the late 1990s, Würbel showed that mice housed under standard laboratory conditions were more likely to develop abnormal, repetitive behaviors known as "stereotypies," such as repeated back flips or gnawing on the cage bars (Würbel, Chapman, and Rutland 1998; Würbel et al. 1996). Based on these findings, Würbel argued that using standard cages might actually compromise the robustness of behavioral experiments by generating abnormal behavioral patterns that would have little meaning outside of the artificial context of the laboratory (Würbel 2000). He argued that researchers should instead systematically vary the cage environment in order to see the full range of variation in a mouse's behavior, rather than attempting to standardize housing conditions for laboratory mice. Würbel's research and his recommendations for alternative laboratory practices challenged the credibility of other researchers' main lines of work, as an article that appeared several years later in the popular science magazine *Discover* made clear: The article, titled "Can We Trust Research Done with Lab Mice?," suggested plainly that scientific experiments conducted using standard cages could be invalid because the animals "may be out of their minds" (Yeoman 2003, 64).

The impact of Würbel's research agenda on that of animal behavior genetics did not go unnoticed by behavior genetics researchers. The inaugural issue of the journal *Genes, Brain, and Behavior* (the official journal of the International Behavioural and Neural Genetics Society (IBANGs), and a favorite publication venue for researchers at Coast), featured a debate between Würbel and two behavioral neuroscientists. In his piece, Würbel critiqued behavioral researchers' practice of recording and reporting on a wide variety of environmental parameters, arguing that this management strategy only created the illusion of control. He wrote that "by pretending 'to list all factors that affect mouse behavior,' such lists may in fact divert attention away from highly relevant factors that were not considered, were considered to be irrelevant, too difficult to assess, or simply cannot be listed" (Würbel 2002, 5). The behavioral neuroscientists' response to this criticism of their by-product management strategies resembled the responses to other criticisms of the animal behavior genetics research agenda that I described in chapter 2: they readily acknowledged that housing and handling conditions could impact behavioral test results and that the present-day lack of knowledge about these factors meant that they were "controlled sub-optimally, if control is not absent at all." However, they hoped that their methodological efforts would put the field on a strong future trajectory, one where the "science will advance and such measures may be developed" (Van Der Staay and Steckler 2002, 10).

THE MANAGEMENT OF BY-PRODUCTS AND THE CREDIBILITY OF GENETIC FINDINGS

The examples in the previous section hint at some of the ways that the stability and credibility of genetic findings is linked to researchers' management of other kinds of knowledge. The tacit/craft knowledge framework emphasizes the importance of this knowledge for getting experiments to work but does not make it easy to see other possible relationships. When thinking of a manufacturing scenario, many interactions between the management of by-products and the value of a company's signature product become evident. The way that a sawmill owner deals with her sawdust might impact everything from the price that she charges for the lumber to the way she chooses to manufacture it. If she treats sawdust as waste, then producing lumber becomes costlier. However, if the market changes and she can sell sawdust as a product, then the cost of producing lumber declines, and she might even be incentivized to mill planks in a way that produces more sawdust. A company might also paradoxically benefit from spending more money to dispose of their by-products (e.g., by treating them to reduce environmental harm prior to disposal), if these efforts enhance the company's reputation and the market value of its other products.

By way of analogy, new relationships between the facts researchers aim to produce and the other kinds of knowledge they accumulate become visible. Just as sawdust is a necessary by-product of lumber production but not something necessarily consigned to the waste bin, researchers have a range of options available to them in managing the knowledge their work creates. Collecting and preserving knowledge with methodological value could offset the time and energy "costs" of producing genetic knowledge, but researchers might also choose to dispose of that knowledge if preserving it decreases the "profitability" of their scientific findings. Researchers might invest heavily in knowing about nongenetic aspects of their work if it enhances their scientific reputation and by extension the value of their genetic facts. They might also use their nongenetic knowledge to effect a "market correction" if they think that other researchers are overvaluing the genetic findings they release into the marketplace. The remainder of this chapter sketches out in more detail four ways that researchers at Coast and beyond used their epistemic by-products to intervene in collective discussions about how to produce knowledge about the genetics of behavior.

Adjusting Epistemic Value

One way that researchers used knowledge about nongenetic differences was to intervene in debates about the power of particular genetic methods. A study published in *Science* magazine in 1999 offers a good example of this technique. In this study, a trio of animal behavior geneticists reported on their efforts to experimentally investigate the extent to which variations in the laboratory environment impacted behavioral test results (Crabbe, Wahlsten, and Dudek 1999). Each of their three laboratories tested eight mouse strains in six different behavioral tests and compared their results. The authors reported that they went to "extraordinary lengths" (1670) to standardize their experimental procedures and laboratory environments. The mice used in the experiment were born on the same day, weaned at the same age, fed the same diet, and slept in the same brand of bedding. Even basic supplies, such as the sheets of sandpaper used for one set of experiments, were shared between laboratories to ensure uniformity, "much to the amusement of the office staff in Edmonton [the location of one of the participating laboratories] who had never seen four sheets of sandpaper delivered by courier" (Wahlsten et al. 2003, 288).

Their results showed that even in the best-case scenario, where the laboratory and housing environments were extraordinarily carefully controlled, the results of some behavioral tests still varied. Some tests, such as the "two bottle choice" assay I will discuss in the next chapter, produced consistent results between laboratories. But in other cases, the differences were quite pronounced: when the researchers injected cocaine into one strain of mouse, the average increase in the mouse's movement in an activity monitor was 701 centimeters per fifteen minutes in Albany, 667 centimeters in Portland, and more than 5,000 centimeters in Edmonton (1672). The study authors concluded by calling for greater caution in interpreting genetic findings, and findings from knockout studies in particular. "For behaviors with smaller genetic effects (such as those likely to characterize most effects of a gene knockout)," they wrote, "there can be important influences of environmental conditions specific to each laboratory, and specific behavioral effects should not be uncritically attributed to genetic manipulations such as targeted gene deletions" (1672).

This public display of knowledge about the impact of the laboratory environment on the behavior of genetically identical mice elicited divergent reactions from their fellow behavioral researchers. For some, the results seemed entirely unsurprising. In a cluster of letters to the editor published in *Science* a few months after the original report, several writers treated these findings and their methodological implications as common knowledge. One letter read, "This important study . . . demonstrates clearly what is widely known in the neuroscience field: behavior is a complex phenomenon that is strongly affected by both genetics and environment. . . . An important message conveyed by this study is that several different approaches should be used, either within or between laboratories, before a definitive interpretation of a behavioral change is made" (Picciotto and Self 1999, 2067). Another letter agreed that the methodological recommendations the authors made were things that researchers should be doing anyway, noting that "[their] own practice [was] never to rely solely on the results of one test, but to apply multiple tests" (Dawson, Flint, and Wilkinson 1999). In a follow-up article discussing their original study, the authors themselves even commented that "no informed scientist should be shocked by a report that environment can alter mouse behavior" (Wahlsten et al. 2003, 306).

But the study authors also noted that not all behavioral researchers came to the same conclusions. Published at the height of debates in the 1990s about the value of knockout techniques for studying behavior, the study authors intended their work to be a pointed but fair cautionary message about the dangers of drawing quick conclusions about complex phenomena. But to others, it was a more radically deflationary move, one that called into question the entire animal behavior genetics enterprise. They wrote in their follow-up piece: "Given our interpretation of the results restated above, we were puzzled when discussing these data publicly to find that many apparently interpreted the results pessimistically to indicate that behavioral tasks were intrinsically unreliable in mice, and that strain differences were unstable across environmental conditions" (Wahlsten et al. 2003, 305).

Dr. Martin recalled the year that this study was published as a "difficult year for [the field]." The study was a hot topic of conversation at neuroscience conferences that year and touched off growing anxieties in the field about failures to replicate the results of knockout studies. Some practitioners saw the study as reinforcing the "old stereotype that many behavioral results are unreliable," as researcher Donald Pfaff (2001, 5957) put it. For Pfaff, the study went too far in highlighting uncertainty, making it look as though behavioral experiments could not generate properly scientific results. Pfaff (2001, 5957) quoted the study in a commentary piece in the *Proceedings of the National Academy of Sciences* where he highlighted some of the "reliable products" coming out of mouse behavior genetics, findings that he argued made "brutally clear" effects of particular gene alterations.

The multisited study and the reactions to it demonstrate how researchers can use epistemic by-products to adjust evaluations of the scientific products that they and others produce. By taking knowledge that was typically discarded or circulated informally among researchers and turning it into a high-profile publication, the authors of the multisited study forced a conversation about the value of animal behavior genetics facts. For those who already shared similar assumptions about the complexity of behavior, the impact of the paper was minimal. However, for others who had previously treated environmental variation as epistemically insignificant, it was a more serious blow to the credibility of their findings.

Although this type of publication may be exceptional, using unpublished knowledge in this way is not. Steve Hilgartner (2017), for example, shows how researchers in genome science routinely traded in gossip and scuttlebutt at conferences, which they used to assess the quality of competitors' results or the trustworthiness of potential collaborators. Sara Delamont and Paul Atkinson (2001) similarly point out that even as new students learn to eliminate evidence of their hard-won craft knowledge from their public presentations, they also learn to use this knowledge to read between the lines and evaluate the practical competences of other laboratories. The epistemic by-products gained through laboratory work can thus serve as a counterbalance to the stylized (or some might say, sanitized) claims that appear in publications and public presentations, allowing researchers to make adjustments to their value.

Altering Experimental Practice

A second way that researchers used their epistemic by-products was to make interventions into their fellow researchers' experimental practice. In the case of the multisited study, the study design left unresolved the question of which environmental factors were responsible for the differences in behavior they observed. Other researchers, however, took this finding as an opening to make more specific recommendations about which features of the laboratory setting researchers should try harder to control. For example, one group of letter writers responding to the study in *Science* magazine pointed to the oft-ignored variation that existed in commercially available mouse chow. They suggested that researchers should switch to more "rigorously defined semi-synthetic diets" in their laboratories (Tordoff et al. 1999, 2069). Another letter writer argued for greater attention to the influence that social rank might have for animals housed in groups (Pohorecky 1999).

Other studies publicizing the effects of the laboratory environment seemed to be designed with the aim of making specific methodological recommendations in mind. Researchers used experimental approaches to transform anecdotal stories about particular environmental effects into data with greater weight for altering experimental practices. A research group in Brazil, for example, conducted a study where they systemically varied several environmental parameters they suspected of causing variation in rodent anxiety tests (Izídio et al. 2005). They found that the position of the animals' home cage (whether it was at the top or the bottom of the shelving units in the colony room) noticeably altered the test scores of some rat strains. Publishing this rather quirky-sounding finding in a respected journal such as *Genes, Brain, and Behavior* may seem exceptional, but this example is far from unique. In their literature review, the authors of this study pointed to more than a dozen other publications describing experiments on other variables suspected of altering anxiety test outcomes, such as age, sex, time of testing, handling, group housing, and transportation to the test room.

In another publication, one research group in Canada took the detailed records they had accumulated from decades of doing pain sensitivity testing on mice in their laboratory and used bioinformatics techniques to turn these records into a source of information on the factors that influenced a mouse's pain sensitivity scores (Chesler et al. 2002a, 2002b). In their analysis, they found that the single biggest variable impacting the results of those pain tests—larger than the genotype of the mouse—was the person who conducted the experiment. Altogether, they found that genotype accounted for only 27 percent of the difference in test scores between mice, while environmental factors accounted for 42 percent of the difference. Other factors that altered their pain test scores in a statistically significant way included the time of day and whether a mouse was the first or last of its cage mates to be tested. By statistically partitioning the variance associated with particular factors, the authors were able to offer specific recommendations about how fellow researchers should alter their laboratory practice. They observed, for example, that even though season and humidity had a detectable influence on test results, "it [was] virtually unheard of for such parameters to be reported in the biobehavioral literature" and that their study suggested "a need for their reporting and/or control" (Chesler et al. 2002a, 918). They also suggested that other research groups could take advantage of their method to conduct statistical analyses on their own records and identify the greatest sources of variation for their own experiments.

Studies such as these are not the core product of animal behavior genetics labs—all of the studies I have described came from groups whose main focus is on genetics or neurobiology—but they should not be discounted as knowledge production anomalies. A comprehensive analysis of the "business model" of the animal behavior genetics laboratory needs to take into account these kinds of products. In the economic logic of the metaphor I have been using, it might seem that animal behavior geneticists have adopted a bad business strategy, one that involves producing the occasional methodological paper at the expense of disclosing their trade secrets or compromising the value of their signature genetic products. But at Coast, these environmental inquiries were not just a side business for researchers—they were an integral part of their overall research agenda. To them, it was those who aimed to produce genetic facts at any cost who had the bad business plan. They believed that researchers who were inattentive to environmental difference increased the likelihood that they would eventually have to revise or even withdraw their findings, doing more harm to their "business" in the long run. The value in doing an experiment on an environmental factor, then, was not solely in the publication that might result, but in the capacity for such research to contribute to the long-term credibility and stability of the animal behavior genetics enterprise.

Managing Scientific Reputations

A third way that researchers used nongenetic knowledge was to establish their reputations as careful scientific practitioners. At Coast, researchers' assumptions about the complexity of behavior made the reputational consequences of dealing with one's epistemic by-products especially apparent. If behavior was a product of both genetic and environmental factors, then handling, holding cages, and dog smells all contributed in a nontrivial way to making the behavior they observed. Under such assumptions, researchers needed to know about and attend to the laboratory environment in particular ways for their genetic facts to be seen as credible—if researchers could not claim an intimate knowledge of both the genetic and environmental conditions from which behaviors emerged, then how could their claims that the differences they observed were due to genetics be trusted? The way that researchers treated their epistemic by-products served as indicators of the quality of the facts that their laboratory produced.

These concerns were evident to me in a public presentation that Dr. Tremblay made to the department on her lab's current research progress. I had spent the week before the presentation shadowing the members of the Tremblay Lab and observing Dr. Tremblay and her graduate students as they sorted through their available data and discussed what was robust enough to present in the talk (or as Dr. Tremblay put it, decide on what seemed "ready for prime time"). Dr. Tremblay decided to present on their work on the genetics of alcohol withdrawal. They had created several lines of mice with variations in a region of the genome that they thought impacted alcohol withdrawal, and they had begun testing those same lines of mice to see how they behaved when withdrawing from other drugs. One of the mouse lines, which had strong symptoms when withdrawing from alcohol, showed almost no symptoms when withdrawing from pentobarbital, a drug used in humans to control convulsions or induce respiratory arrest in euthanasia. The difference in the mice's response to those two drugs was surprising. Given the similarities in the way that alcohol and sedative hypnotics affect the body, Dr. Tremblay had expected that the patterns of withdrawal symptoms would be very similar. This unexpected finding was now guiding their research agenda, and they had recently begun testing other mouse lines and other drugs to see if they could find more such differences.

After the talk, Dr. Tremblay and several of her graduate students clustered in the hallway just outside of the conference room, analyzing their colleagues' responses. I stopped to chat with them, but they were so involved in conversation that they barely acknowledged my presence. A question of Dr. Smith's had them worried—he had asked about the technique they had used to make the mice addicted to the drugs before testing them for withdrawal and suggested that the particular way they had administered the pentobarbital to the mice may have been the reason they did not see the expected withdrawal symptoms. As Dr. Tremblay and her grad students discussed this possibility, they also ran through other differences between the alcohol and the pentobarbital testing procedures and realized that all of the alcohol testing had been conducted by one person and all of the pentobarbital testing by another. The difference they had been seeing, then, might be related to the individual researchers' technique or smell, and not to the unique genetic composition of that line. This failure to investigate and control for possible impacts of the experimenter on withdrawal behavior was, in their minds, something that called into question the entire line of research.

Not all researchers would have concluded that the difference between experimenters was consequential enough to cast doubt on their withdrawal experiments. While at Coast it was taken as a given that the experimenter could substantially impact mouse behavior, others might have concluded that such a difference was unlikely to be a cause for concern. This was one of the points of contention in the knockout debates of the 1990s, with those trained in molecular biology arguing that "small" differences between experiments were unlikely to overshadow the "large" effects of eliminating a gene, while behaviorists argued the opposite.[15] In the end, the Tremblay Laboratory decided to redo both the alcohol and the pentobarbital withdrawal studies with the same person conducting all of the experiments. Those studies might have revealed that their promising results were in fact only experimenter effects, but choosing to ignore this factor might have done even more damage to the reputation of the laboratory by suggesting that they were not careful practitioners.

Grounding Experimental Facts

A final way that Coast researchers used knowledge by-products was to situate genetic facts in the circumstances of their production, making them seem less universal than they otherwise might. In the introductory behavior genetics class at Coast, Dr. Martin opened one of her lectures with a story about a set of experiments conducted in the late 1990s in a laboratory that she was working in. Her research group was testing a knockout mouse missing a gene that they thought was related to anxiety. So, they ran the mouse through three standard tests for anxiety-like behavior: the elevated plus maze, the light–dark test, and the open field test. As they were concluding their experiments and preparing the data for publication, they discovered that two other laboratories had also created knockout mice missing the same gene and had even used the same anxiety tests. The three research groups agreed to coordinate their publication efforts and report them together in *Nature Genetics*. But when they began to compare their studies, they found their results differed in noticeable ways. While one group reported that the knockout mouse had the same level of anxiety as the unaltered mouse in all three tests, the second group found that the knockout mouse was less anxious than the wild type in the same tests, and the final group found that the knockout was less anxious in two tests but more anxious in a third.

The researchers' reaction to these contradictory results departed substantially from what we might expect based on existing STS studies of scientific controversies. The researchers did not attempt to arrive at a consensus about what constituted a well-done knockout experiment or to discount one laboratory's results through a detailed critique of their experimental methodology. Dr. Martin recalled that the editor of the journal contacted them and asked them to consult with one another so that they could "come to agreement" about what kinds of conclusions they could make about the role of this gene in anxiety before publication. But the editor's attempt to facilitate a process of social negotiation among the three research groups was met with surprise and even some indignation by the researchers in her laboratory. Dr. Martin recalled that she felt that they were being asked to "massage" their results. Rather than looking for a single correct interpretation, Dr. Martin's opinion was that all of the results might have some truth to them. She recalled that when she looked at the papers from the other two laboratories, her first impression was that differences in the testing protocols alone were enough to explain why the other groups had different findings. The light levels in the three laboratories, for example, were markedly different, an environmental factor that she immediately recognized as having a substantial impact on the manifestation of anxiety-like behavior. In the end, the research groups agreed to publish all three studies with their conflicting results in place and with notes in experimental reports that such discrepancies might be due to differences in either the way that the knockouts were constructed or in the test environment.

This example shows how the knowledge gained through the process of behavioral experimentation can ground genetic facts, embedding them in their circumstances of production. As a result of the authors' insistence on preserving difference, what emerged from these knockout experiments was not a single, stable scientific fact, disconnected from the people and places that produced it. Rather, it was a set of divergent results that were each tied to a specific, local set of experimental procedures. It is an example of how locality and uncertainty might be inscribed into "immutable mobiles" (Latour 1987) rather than erased, preserved in print for all to see. What is especially notable is how the researchers involved in this episode treated divergent and locally confined findings as the normal state of affairs, rather than an aberration. At least in Dr. Martin's retelling of the story, no one was surprised to find that their test results were different—what surprised them was the journal editor's insistence that they should arrive at a single conclusion. One of the three papers stated simply that "it [was] known" that elevated plus maze results can vary with environmental changes, deploying a body of knowledge for which no citation was provided but that the authors presumed was commonly held. And rather than working to eliminate contradictory information and resolve multiplicity into stable findings, the researchers used their epistemic by-products to explain and preserve the difference in their experimental findings.

CONCLUSION

The collective dynamics that I have described in this chapter—where researchers affect reevaluations and rereadings of one another's knowledge—might feel familiar to STS readers, because this is one of the ways that we as analysts engage with science. Like the infrastructure of the Mouse Phenome Project or the alternative perspectives of welfare-oriented researchers, STS analysts also collect up scattered observations so that new patterns can emerge and draw attention to aspects of laboratory practice that practitioners treat as unimportant. Analysts often read scientific papers against the grain, sometimes with the explicit aim of reevaluating scientific findings. For example, Vinciane Despret (2004) examines psychologist Robert Rosenthal's (1966) classic work on experimenter effects with the aim of developing new theories of how animals and experimenters affect each other in the laboratory setting. This approach, she notes, involves an inversion of the value that Rosenthal attributed to his own experimental observations. She writes: "As a matter of fact, the study of these 'little differences' that Rosenthal wanted to spot, these differences that affect the subject making him or her respond differently, was a marvellous idea. But Rosenthal's original idea had not been to explore a world enriched and created by these differences; it had been to mark them off as parasitic supplements that seriously contaminate the purity of the experiment" (Despret 2004, 118). Using my terminology, we might say that psychology's by-products are Despret's source materials for reconceptualizing animal-human relations. The things Rosenthal treated as observations collected in the service of furthering psychology's main epistemic agenda, Despret treats as important products in their own right.

Joan Fujimura (2006) has articulated a similar project of attempting to reread data that scientists have discounted or ignored. She argues that laboratory work produces an "awkward surplus" of data that does not fit with scientists' preconceptions about their subject matter and that analysts can reinterpret using different frames of reference to reach alternative conclusions. In attending to this surplus, Fujimura aims to open up research on sex differences to a broader set of concerns. She explicitly compares the role of social science analysts and social activists to that of scientists from other fields, who could "see anomalies as sources of novel ideas and findings because they bring different assumptions to the table" (70). Involving other groups in sex research, she argues, could create opportunities for the scientists conducting that research to think more creatively about what sex is.

However, the reasons why scientists themselves might engage in these activities or preserve the material that analysts use for their rereadings are less clear. Why would Rosenthal be interested in revealing the uncertainties and contingencies in his field's own knowledge production processes? And why would other psychologists praise his work, rather than ignoring it or attempting to discredit it?[16] Similarly, why would researchers include information on cases that did not fit their theories of sex determination in their publications? There is no shortage of examples where researchers have simply omitted data points they believed to be anomalous,[17] so if Fujimura's sex researchers believed that cases of indeterminate sex were unimportant, why would they use their limited publication space to describe them?

Making this work understandable requires moving outside of some existing analytical frameworks in STS. Once again, an individualistic, competitive view of laboratory life is limited in its capacity to explain these dynamics. It might make sense that researchers might use their laboratory experience to undermine a competitor's results, but a competitive framework offers little explanation of why researchers might cast doubt on the stability of their own findings. Likewise, it is tempting to describe knowledge about dog smells or the position of the home cage on a cage rack as tacit or craft knowledge, because it often circulates in the informal, face-to-face channels that have been so well described in previous studies. But these categories make it harder to appreciate how readily knowledge about the environment can be formalized and published, and why researchers might want to do so. My goal is not to weigh in on questions about whether there really exists a kind of knowledge that is only communicable through face-to-face interactions. Instead, I aim to shake up the many empirical observations that have settled into the category of tacit/craft knowledge to make them available for new kinds of theorizing.

Taking into account extrafactual activities such as scaffold work and by-product management provides a more comprehensive picture of what the laboratory produces. From the perspective of the scientific literature, it would be easy to see animal behavior geneticists as focused on genetics and only minimally concerned with the environment. If you read a publication of Dr. Tremblay's, for example, it might seem reasonable to conclude that her laboratory paid little attention the nongenetic factors influencing the severity of alcohol withdrawal and perhaps that she believed such a behavior to be largely genetic. But this would be like trying to evaluate a lumber company by looking only at a single product, without knowing anything about its costs, production processes, or long-term business strategy. When taking these other processes into account, we might see animal behavior geneticists not as spending tremendous energy to produce relatively few genetic facts, but as investing effort in building conceptual and technical foundations for future research.

CHAPTER 5

Understanding Binge Drinking

Taped to the wall next to the door of a laboratory at Coast University is a copy of a photograph showing a middle-aged man on a street corner (figure 6). The man is sitting on an overturned milk crate, with an empty McDonald's cup placed in front of him. His heavy blue jacket and black pants blend in with the gray street scene, and his face is turned away from the camera so that only the back of his baseball cap and his ponytail are visible. In the center of the photograph, the man is holding a sign with the message "Need cash for alcohol research" written in black marker. Underneath the photograph, someone has drawn an arrow pointing to the man in the picture and added a handwritten note that jokes, "When Dr. Smith had a ponytail."

Who is the anonymous man in the picture? A witty panhandler? An alcoholic? Or, as someone at Coast has jokingly suggested, a fellow alcohol researcher in need of funding? The image plays on cultural stereotypes about who the alcoholic is: an older man, alone, without a job, and possibly homeless whose day is organized around getting alcohol or money for more alcohol. This picture may also conjure up ideas of how the alcoholic came to be this way and how society should respond. Some might see a man who has chosen a destructive lifestyle that has isolated him from friends and family, and others might see a man who is suffering from the disease of alcoholism. Perhaps his present situation was exacerbated by a genetic predisposition to alcoholism, or stressful life events, or by poverty and inadequate housing. Maybe the man in the image needs to take responsibility for his drinking, or admit that he has no control over his consumption of alcohol. Maybe new medical interventions

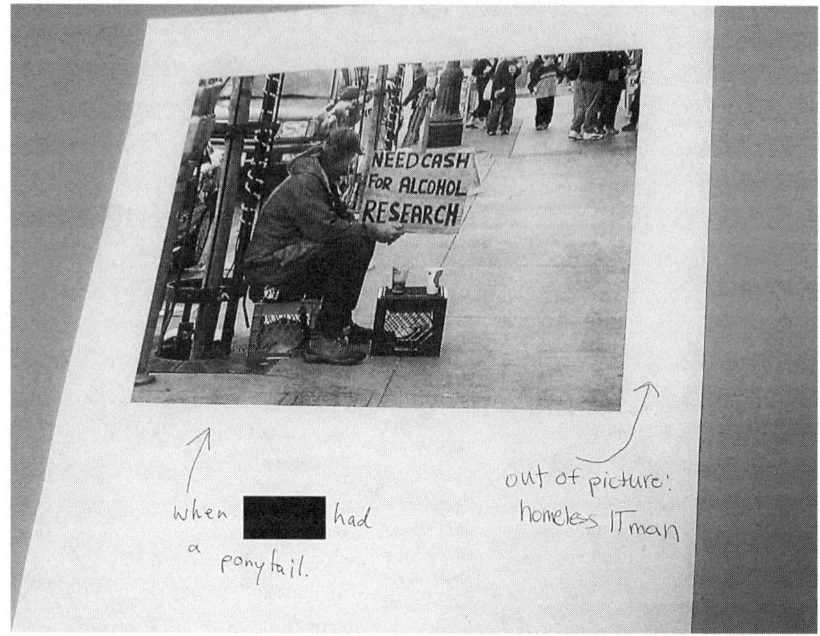

FIGURE 6. Image posted in the hallway, with handwritten captions presumably added by graduate students. Photograph by the author, April 2008.

or public health programs are needed to help him stop drinking, or to keep others like him from drinking in the first place.

Commentators from both inside and outside of the sciences have been critical of how biomedical research shapes cultural understandings of drinking. While a photo of a man on the street in search of alcohol may lend itself to a variety of thoughts about the causes of and solutions to problem drinking, critics express concern that biomedical research promotes particular "governing images" (Room 1974) that encourage us to see drinking in some ways and not others. Specifically, critics argue that biomedical research promotes biologically reductionist and/or determinist understandings of drinking, and obscures the structural, cultural, and historical forces shaping these behaviors. These criticisms present a paradox for understanding research at Coast: although researchers expressed commitments to complexity, they worked with genetic animal models—an arguably reductionist approach to understanding addiction. Researchers are subject to structural pressures of their own (such as the strongly biomedical orientation of funding agencies such as the National Institute on Alcohol Abuse and Alcoholism [NIAAA] and the National Institute on Drug Abuse [NIDA]); but even so, why would a researcher who describes addiction as a complex, multifactorial phenomenon choose to study it using a mouse in a cage?

This chapter explores the apparent disconnect between researchers' theories and their practices by examining the development of a new mouse model of binge drinking. The model was developed within the context of an interdisciplinary Alcohol Research Group (ARG) funded by the NIAAA as part of an effort to accelerate research in understudied aspects of alcoholism. The ARG researchers developed a model that commentators might argue promotes simplified, biologized understandings of complex biopsychosocial problems. The researchers themselves, however, saw their work as grounded in an understanding of binge drinking as a complex behavior. This contradiction lessens when we take into account the aspects of experimental life that I have discussed throughout this book. Considering the epistemic by-products that researchers accumulated as they built this new test, for example, makes its interpretive flexibility more evident. For researchers who spent many months adjusting a mouse's home cage environment to encourage it to drink more, the test was as much a model of the situational triggers leading to binge drinking as it was a model of genetic susceptibility. I argue that understanding this interpretive flexibility is useful for thinking about how to engage in critical dialogue with those developing animal models of addiction.

INTERDISCIPLINARY RESEARCH IN THE ARG

In the early 2000s, the NIAAA funded several large interdisciplinary projects that I will refer to as ARGs. These projects were a new type of funding arrangement within the NIAAA, designed to bring together alcohol researchers with different disciplinary backgrounds to study aspects of the same model, and to recruit new researchers to the alcohol field. Researchers were awarded a package of research grants as a group to focus on one particular problem and were encouraged to bring a variety of different approaches and techniques together to study that problem. According to an early progress report, these projects represented "the largest ever concerted effort to collect and integrate scientific data on neuroadaptation to alcohol consumption," using an approach that "combined study of animal behavioral models with molecular, cellular and systems-level measures of brain function."

The NIAAA proposed this novel funding arrangement to solve a particular knowledge production problem: different research groups tended to use different animal models of alcoholism, which made it difficult to integrate their data. Dr. Raymond Williams, a senior NIAAA official, recalled that he first began to recognize this problem in the mid-1990s when he was undertaking a comprehensive evaluation of the neuroscience projects funded by the institute. He said,

> One thing that became apparent, at least to me, was the fact that people were working with their little unit in terms of research. So if they had one model, people would work on that model. Another guy would work on the same model but change things a bit, both of them would modify something, and if you looked at the data, the results coming out, you could never correlate them, even say supposing you were looking at something as simple as a two bottle choice, OK? Some people do twenty-four hours, some people do limited access, you know different schedules, and you could never correlate one or the other. And you know, pieces of data were coming out and we were not confident that we could make a legit statement.

Even the relatively straightforward "two bottle choice" experiment—where researchers give mice a bottle of water and a bottle of alcohol solution and measure how much they drink from each—was subject to these variations: While one researcher might leave both bottles on the cages for a full twenty-four-hour period, others might run the two bottle choice procedure by giving mice access to water and alcohol for only part of the day. These kinds of variations complicated comparisons across laboratories, because researchers regarded these different experimental procedures as to some extent generating different behaviors, or at least as tapping into different aspects of a larger behavior.

The NIAAA's aim was to provide researchers with an incentive to focus in on a single, standardized model. And rather than just bringing mouse researchers together around a shared protocol, they adopted the more ambitious strategy of bringing multiple scientific approaches into conversation with one another. The ARG program encouraged researchers to develop models that could work across a variety of experimental organisms and human populations and to use the techniques of many different disciplinary specialties to study that model. A single ARG, then, might include behavioral researchers working with mice, rats, flies, monkeys, and humans, as well as molecular biologists, electrophysiologists, neuroanatomists, and bioinformaticians. The degree to which these initiatives were successful in generating consensus around a single model is, as we will see, debatable. But they did create a venue where researchers from a variety of backgrounds could talk explicitly about methodology in animal modeling and the relationship between animal models and human disorders.

Members of the neuroscience department at Coast were involved in several of these initiatives, and I followed one group who had decided to focus their collective efforts on modeling "binge drinking" behavior—a behavior the group agreed was both a pressing public problem and an underdeveloped area in rodent research. The NIAAA (2004) defines binge drinking in humans as "a pattern of drinking alcohol that brings blood alcohol concentration (BAC) to 0.08 gram percent or above. This typically happens when men consume 5 or more drinks, and when women consume 4 or more drinks, in about 2 hours." Researchers thought it was important to study this type of drinking behavior because it is a risk factor for developing alcohol dependence and also causes health and social problems in its own right. People who drink in binges, for example, are fourteen times more likely to report that they have driven while under the influence of alcohol compared to nonbinge drinkers (Naimi et al. 2003). It is also a very prevalent phenomenon in the United States: some research estimates that approximately 75 percent of the alcohol consumed by American adults is consumed in "binges," and this figure may be as high as 90 percent for adults under the age of twenty-one (Underage Drinking Enforcement Training Center 2005).

Relatively little research existed on binge drinking, however, because of a property of mouse behavior and biology that was well known in the field: mice do not like to drink. As Edmund Ramsden (2015) has explored, this property of mouse behavior has long complicated researchers' efforts to develop animal models of alcoholism. While mice will self-administer morphine or cocaine, when given a bottle of alcohol solution only a few strains of mice will drink a good quantity, most will drink only a little, and some will drink none at all. Ramsden (2015) argues that researchers working at the Center of Alcohol Studies in the 1970s were acutely aware of that the lack of an animal model that would drink heavily created a credibility problem for the alcohol field, and listed the development of such a model as one of their main objectives. Alcohol researchers have offered a variety of reasons to explain why rodent models of self-administration have been so difficult to develop. Mice are "neophobic"—they are fearful in general of new things and new foods in particular, because they have no vomiting reflex and cannot purge something once they have ingested it. Alcohol tastes and smells bad (perhaps especially so to certain strains of mice), and mice might avoid it because of these "adverse orosensory properties." Mice are prey animals and may avoid getting intoxicated because it could decrease their ability to detect and escape from predators. Some researchers have suggested that mice might not even be capable of "drinking to intoxication" because they metabolize alcohol so quickly and would therefore have to consume impossibly large quantities in order to actually get drunk.

Alcohol researchers have developed a variety of techniques for "unmaking" mice as reluctant drinkers and "remaking" them into the kinds of bodies that could substitute for those of human binge drinkers, to use Mette Svedsen and Lene Koch's (2013) vocabulary. A few of the techniques that are known to increase the amount that mice will drink include sweetening and flavoring alcohol solutions, offering food along with the alcohol, or providing alcohol bottles at regularly scheduled times of the day. Researchers also employ more direct approaches to get alcohol into the bodies of mice, such as housing them in specially designed chambers where they are forced to breathe in alcohol vapor or injecting alcohol directly into their abdomens. But for different reasons, the ARG researchers were dissatisfied with all of these existing techniques. To take one example: abdominal injections are easy to do and ensure that every mouse receives the same dose of alcohol, but the ARG researchers thought that this expediency came at the expense of the "face validity" of the model. As one researcher put it for me, injections were a "tough sell" as a model for a weekend bender because "nobody sits around doing IV alcohol, except for maybe Hunter S. Thompson." What the researchers decided they needed, then, was a new model.

The model that the ARG researchers eventually settled on I will call the "nocturnal drinking" model. This protocol used existing knowledge about patterns in the circadian rhythms of mice to encourage heavy drinking. As nocturnal creatures, mice tend to be most active a few hours after waking up at night. The ARG researchers speculated that by offering alcohol to mice at this time of day, when they already tended to consume most of their food and water, they would drink more alcohol. The protocol they proposed was straightforward: C57/B6 mice (who were known already to have a preference for alcohol) were kept in standard cages with food and water bottles, but a few hours after the lights went off in the housing rooms, researchers would switch the water bottle for a bottle of alcohol. And so for that short window of the day when mice were most inclined to eat and drink, they would have only a bottle of alcohol available to them. The protocol also fit conveniently within the rhythms of the laboratory; by housing mice in "reverse light-dark rooms," where the lights are on at night and off during the day, researchers could flip the circadian rhythms of their mice and perform nocturnal drinking experiments during normal working hours. Researchers would switch the water bottles with alcohol bottles when they arrived in the lab, and a few hours later they could check the mice to observe how their behavior and biology changed as they drank.

Most importantly, the protocol was successful in fulfilling what the group had collectively defined as the key criterion for evaluating their proposed animal models of binge drinking: the mice drank enough to reach a BAC of 0.08 gram percent, the same as the NIAAA's definition of binge drinking. Researchers could take a blood sample a few hours after they had put the bottles of alcohol on the cages and show that the mice were reaching BACs of 0.08, 0.09, or even 0.10 on average.[1] They could also run the mice through standard behavioral tests, such as observing their footsteps as they walked a mouse-sized balance beam, to show that the mice also showed the symptoms of intoxication. This evidence, along with the practical features of the protocol, made it an attractive model for researchers in the ARG and beyond to adopt. It required no long training periods to acclimate the mice to alcohol or expensive equipment such as vapor chambers. It was supported by pharmacological and behavioral evidence and remedied what many saw as a critical weakness in the epistemic scaffolds of existing models. Within a decade, numerous research groups and even some pharmaceutical companies were using the nocturnal drinking model.[2] Even Dr. Smith seemed unusually enthusiastic. He commented to me that he found it easier to discuss this particular project with the news media, because he thought it was a clear case of a scientific advance. "This is a better model," he told me, "because the mice really do drink and fall over just like college students."

NOCTURNAL DRINKING AS A REDUCTIONIST MODEL?

This model—a lone mouse drinking in its cage, judged for its ability to model human behavior by its BAC, and designed to facilitate genetic and neuroscientific studies of binge drinking—represents an approach to studying human problems that analysts have been critiquing for at least thirty years. From the elimination of social and structural considerations from the experimental context to the (bio)medicalization of deviant behavior, the nocturnal drinking model touches on many well-worn arguments about the limitations and dangers of these approaches.

Alcoholism exemplifies what scholars have described as medicalization: the pervasive tendency to recast behaviors that violate social norms as biological problems requiring medical intervention. As much scholarship in this area has argued, the conception of alcoholism as a biological, genetic, or a "brain-based" illness is culturally and historically specific (Campbell 2007, 2010; Conrad and Schneider 1980; Levine 1978; Room 1983; Schneider 1978; Vrecko 2010a, 2010b). Conrad and Schneider (1980) have famously argued that while heavy drinking may once have been perceived primarily as evidence of moral failing, it is now largely viewed as an illness to be treated by self-help groups and medical professionals. Robin Room (1983) succinctly described this view as the "disease concept of alcoholism," named after the title of a book by alcohol researcher E. M. Jellinek (1960).[3]

Analysts studying drugs and society have taken issue with this conception because they argue it flattens a complicated picture of causation, focusing attention on biology and pushing structural causes out of the frame.[4] In so doing, it discourages investigations into other potentially useful ways of explaining or intervening on addiction (see, e.g., Acker 2010; Hall, Carter, and Forlini 2015) and encourages drinkers to adopt a fatalistic attitude toward their behaviors and prospects for changing them (Fingarette 1988; Kvaale, Haslam, and Gottdiener 2013; Phelan 2005). Scholars have made similar critiques of genetic research more generally.[5] As Steven Rose (1997, 297) has put it, treating social problems as biomedical problems means "attention and funding are diverted from the social to the molecular." "If the streets of Moscow are full of vodka-soaked drunks," he writes, "the ideology [of reductionism] demands the funding of research into the genetics and biochemistry of alcoholism" (297).

Scholarship on animals in science suggests that modeling human disorders using animals can intensify processes of biological reductionism. Nancy Campbell (2007) argues that animal models, in particular nonhuman primate models, played a central role in the turn toward a neurophysiological conception of addiction in the interwar period in the United States. By using monkeys as experimental subjects, she argues that researchers were able to bracket out questions about the role of desire and craving in addiction and study addiction in terms of observable behavior and physiological responses. This strategy allowed them to circumvent the psychoanalytic framings that permeated human addiction research during this period, and to place their research on the "more objective ground" sought by policy makers (Campbell 2007, 29). Nikolas Rose and Joelle Abi-Rached (2013) argue that animal models were "crucial" to the rise of neuroscience more generally. They write that researchers' ability to link psychiatric drugs, molecular information, and behavior through animal models "almost inescapably led to the belief that the anomalies in those mental states could and should be understood in terms of specific disturbances, disruptions, or malfunctions in neuromolecular processes" (10). They are critical of this turn, questioning whether animal models can capture the "rich meaningful, culturally embedded, historically shaped, linguistically organized, situationally framed, *experience* of depression, anxiety, and whatever else in our everyday human world" (83). Donna Haraway's (1989) classic study of primatology points to additional routes through which animal research might lead to the biologization of human behavior. She shows how primatologists constructed nature/culture binaries that placed the raced and sexed narratives they told about primates on the side of the natural world, inscribing stories about behavior and social order into our shared biology.

Joseph Gusfield's (1981) study of drunk driving laws illustrates how such biologized frameworks for understanding behavior can result in problematic public policy. Gusfield explores how in the context of increasing public concern about alcohol and highway accidents, laws regulating the blood alcohol levels of drivers came to be seen as the appropriate solution to this problem. Establishing BAC thresholds allowed policy makers and law enforcement officers to cleanly differentiate those who were "drunk drivers" from those who were not, but Gusfield shows how this solution was a partial one in many ways. For one, blood alcohol levels are not isometric with impairment, and impairment is not a problem that is exclusive to drinkers alone. Laws structured around BAC levels do not address other factors that might predispose individuals to accidents, such as sleep deprivation, or behaviors causing highway accidents that are not exclusive to those driving under the influence of alcohol (such as speeding or tailgating). BAC laws also direct attention away from potential structural solutions to the problem, such as policies to discourage drinkers from driving to bars, training bartenders to recognize signs of inebriation, or technological solutions to prevent impaired drivers from operating vehicles. Much like the ARG's nocturnal drinking model, drunk driving laws seem to solve the problem of impaired driving by only respecifying the problem of impairment as a problem of blood alcohol levels.

Given the sheer volume of critical scholarship on addiction, genetics, and animal models, much of it extending back decades, it is easy to see how an analyst might look at the nocturnal drinking model and see it as an extension of a long line of biologically reductionist research on human behavior. Along with that diagnosis might come a question: If the researchers at Coast and in the ARG really believed in the complexity of behavior, then why would they develop a model that looks like this? That researchers designed a protocol around a mouse locked in a cage with an alcohol bottle when given a multiyear grant to imagine a new animal model seems to suggest either a rather large failure of imagination or perhaps that their complexity talk was just talk.

This chapter aims to explain how research programs such as the nocturnal drinking model could look so different from different vantage points and especially from the vantage points of analysts and practitioners. My aim here is not to make a determination on whether the nocturnal drinking model is *really* reductionist or whether the ARG practitioners *really* believed in complexity—as I will discuss more later, I think that these questions are problematic because there are no independent positions from which to make such evaluations. Nor am I aiming for a reparative reading of this binge drinking model that might bring biological research into greater contact with history and social theory (although I think that there is great value in such readings). What I want to understand is how this particular model could be flexible enough to strike some as obviously reductionist and others as perfectly compatible with a complex understanding of behavior. The key to making this more comprehensible, I argue, is to examine researchers' extrafactual work. It is through tracing the process of developing the foundations of these models that distinctions between researchers' ontological and epistemological commitments become visible and the importance of the epistemic by-products to researchers' understandings of mouse and human behavior become clear.

BUILDING THE EPISTEMIC SCAFFOLD OF THE NOCTURNAL DRINKING MODEL

As we have seen already, in building the epistemic scaffold of their new model, the ARG researchers chose to focus on BAC as a key foundational link between mouse and human behavior. This was a powerful foundation from which to build a new model because of the scientific and cultural significance of the 0.08 gram percent threshold. The NIAAA's own definition of binge drinking relied on this threshold, lending legitimacy to this criterion. It was a standard that could be applied equally to humans and a wide variety of experimental organisms—a feature that was especially attractive given the highly interdisciplinary composition of the consortium. High blood alcohol levels had also long been considered a missing link in the scaffolds of existing models. Demonstrating that mice could be induced to drink to an intoxicating BAC was both a technical and a theoretical advance for the alcohol research field, because many practitioners thought that such an achievement was biologically or practically impossible. Dr. Larry Wilson, a senior behavioral researcher and one of the founding members of the ARG, recalled that even senior members of the group held this view and raised objections about the feasibility of setting a BAC target in their initial discussions. In Dr. Wilson's view, overturning the "old story" that mice could not "drink past metabolism" and demonstrating that they would drink enough to cross the blood alcohol threshold was one of the major success stories of the ARG's work.

In addition to these scientific reasons for focusing on blood alcohol, their choice of the 0.08 gram percent threshold aligned nicely with existing established legal, medical, and cultural boundaries between acceptable and problem drinking. All fifty states in the United States recognize driving with a 0.08 percent blood alcohol as illegal—drivers operating a motor vehicle over this threshold can be charged with a criminal offense whether or not they meet behavioral definitions of impairment. BAC thresholds are also cultural as well as legal objects, and the ARG researchers used established cultural narratives about problem drinking and BAC to reinforce the face validity of their new model. When talking informally among themselves and to me, they often claimed to have generated mice that would "blow over" and "couldn't drive home."

Despite the strength of this foundational link, however, it proved not robust enough on its own to support the nocturnal model as a knowledge production tool. The ARG researchers produced several other kinds of evidence in addition to BAC measurements to support the new model, and even still, they encountered skepticism about the knowledge production capacities of the model. The ARG researchers did experiments to show that mice in this model had impaired motor coordination, suggesting that not only did the mice have similar pharmacological exposure to alcohol as binge drinking humans, but they also showed similar behavioral patterns. The ARG researchers emphasized that the mice in this model were "self-administering" alcohol "via the oral route," both of which satisfied criteria in an established framework for evaluating the validity of animal models of alcoholism, known as "Cicero's criteria" in the field.[6] They also pointed out the practical advantages of this model over existing models, such as the lack of a time-consuming training period.

But even with this additional support, there were still researchers even inside the ARG who doubted that this new model could produce meaningful information about binge drinking. Dr. Frank White, an ARG member, was one of the most vocal critics of the new model. He thought it was "kind of a disaster" because the mice had no choice but to drink alcohol when they were eating because it was the only available source of fluid. He told me that he thought there had been a kind of "radicalization" within the ARG, where the scientific directors had committed themselves to producing high BACs in their animals at the cost of the validity of the models.

The question of what was motivating mice to drink was Dr. White's main concern.[7] The mice in the nocturnal drinking paradigm may have been drinking a lot, but he argued that researchers needed to know more about *why* they were drinking to consider it a useful model of human binge drinking. A mouse that drank simply because it was thirsty or hungry was not useful because it was not drinking for the same reason a human would—the intoxicating "pharmacological effect" of alcohol. He told me, "My question would be, are [the mice] just selected to drink a lot during a certain period, and they don't even know whether it's alcohol or not, and they don't care? They don't have enough experience with alcohol to be able to associate its taste with its pharmacological effects. So, are you looking at the same neurobiological mechanisms in this animal that you would be in a free choice model, where animals have a little more time over a couple of weeks or month to familiarize themselves with the taste of alcohol?"

Dr. White worried that the nocturnal drinking protocol was too short for the mice to be able to learn to associate drinking alcohol with the bodily sensations that it produced. While the short testing period might have had practical advantages for laboratory work, in Dr. White's view it fell short of providing a scenario in which mice arguably knew what they could expect when they drank from the alcohol bottle. The mere fact that mice were ingesting enough alcohol to reach a high BAC was not enough to make it a useful tool for producing knowledge about human drinkers, he argued, because the genetic and molecular mechanisms underlying that kind of simplistic drinking behavior might not be same as the ones associated with the desire to feel drunk.

Motivation was a concern for another ARG member, senior mouse researcher Dr. David James. He also thought that the nocturnal drinking model was problematic because it failed to separate the motivation to experience intoxication from other potential reasons the animals might be drinking. He told me:

DJ: I have concerns that a lot of the drinking models aren't necessarily studying drinking that's motivated by intoxicating pharmacological effects.

NCN: What do you think it's being motivated by?

DJ: Taste, calories? I think that [C57 mice] are drinking because it tastes sweet to them. I think there's evidence that it's affecting sweet receptors, and I don't think they're as affected by the adverse pharmacological effects.

Like Dr. White, Dr. James was worried that without establishing why mice were drinking in the nocturnal model, future researchers risked identifying genes or brain regions involved in taste perception rather than susceptibility to addiction.

Perhaps the most striking evidence of the general preoccupation with motivation in the ARG is to be found in the model that the group developed and then rejected, called the "limited water" model. In this protocol, researchers put water bottles on the mouse cages for only ten hours a day instead of making water available to the mice at all times. During the other fourteen hours when the mice had no water available to them, researchers gave the mice a bottle of alcohol for half an hour. Initial studies showed that under these conditions, the mice would drink substantial amounts of alcohol in the half-hour window that the bottles were on their cages, generating BACs of 0.10 gram percent or higher. The nocturnal drinking and limited water models were developed alongside each other (with the same senior members of the ARG involved in both projects), unveiled simultaneously at one of the annual ARG meetings, and published within a few months of each other. But while the nocturnal drinking model flourished, the limited water model faltered. Only a year and a half after it was introduced, the ARG decided to drop the model from their future research plans. Ten years after the publication of the initial papers, the nocturnal drinking model had accrued six times as many citations as the limited water model.

In many ways, the two protocols were quite similar. Both created situations where animals would drink alcohol in solution and were successful in getting them to drink enough to reach 0.08 percent BACs or more. Both involved removing the water bottles from the cages and substituting them for bottles of alcohol. And in both protocols, this substitution was made on a regular schedule and the alcohol bottles were left on the cage for a relatively short period of time, which meant that researchers could schedule other experiments around the drinking window. But whereas the nocturnal drinking protocol allowed mice to have as much water as they wanted outside of the drinking window, the limited water model required that researchers gradually reduce the time that mice had access to the water bottle in order to get them to drink heavily. Researchers worried that using fluid deprivation in the limited water model created a "motivational confound," where it was no longer possible to tell if the mice were drinking because they enjoyed the effects of alcohol or simply because they were thirsty. Researchers were also suspicious that the drinking taking place in this scenario was not analogous to human drinking because the mice would not drink as heavily if they had water available to them for more than ten hours a day. As I have outlined already, researchers within the ARG argued that if they started doing genetic and neurobiological research using this model, they might end up mapping the biology of thirst or fluid intake rather than the biology of binge drinking. This issue came up both within conversations in the ARG and in the comments from the external reviewers during one of the project renewals, and according to the group leader, concerns about motivation were one of the main reasons why the model was defunded.[8]

Through sketching out discussions around the epistemic scaffolds supporting the nocturnal drinking model, it becomes clear that BAC levels were not the only factor under consideration. Modeling the motivations of human drinkers was as important—if not more important—than the presence of a pharmacological agent in the body. As we will see next, a few ARG researchers were even willing to forgo high blood alcohol levels to achieve greater clarity on whether their animals were drinking for "pharmacological effect." While it might appear to those outside the alcohol field that the nocturnal drinking model was designed with a biologically reductionist definition of binge drinking in mind, examining the group's work suggests that the ARG researchers were using several other implicit criteria to evaluate their models. If BAC was truly the only criterion that researchers cared about, then the limited water model would have been the better choice. But for both the ARG researchers and their colleagues outside of the research group, BAC alone was not enough to convince them that these models could generate valuable knowledge about human drinking.

ALTERNATE MODELS AND EPISTEMOLOGICAL CONCERNS

In part due to concerns about the validity of the nocturnal drinking model, dissenting researchers in the ARG began developing their own new models. Dr. James, who was concerned about taste, developed a complicated experimental setup that allowed mice to "drink" and feel the pharmacological effects of alcohol but without actually smelling or tasting it. He surgically implanted feeding tubes into the stomachs of his mice, and then placed them in cages with bottles containing two different flavors of water. When a mouse licked the cherry-flavored bottle, water would be pumped into its stomach; when it licked the grape-flavored bottle, it would get a dose of alcohol through the surgically implanted tube. In this way, his mice could "drink" alcohol by licking the grape-flavored bottle but without experiencing the burning sensation of alcohol as it passed down their throats.

These new experimental conditions generated some surprising results. Using this experimental setup, Dr. James tested both C57 mice, known as reliable drinkers, and DBA mice, known as the "teetotalers" of the mouse world. In this new model, the C57 mice behaved much as they always had, taking small "sips" from the grape-flavored bottle that allowed them to drink throughout the day but without actually getting very drunk. But the DBA mice's behavior underwent a rather dramatic reversal: not only did the DBA mice suddenly drink more than the C57 mice, but they also drank large quantities all at once, earning them the nickname of the "gulpers." Dr. James argued that his protocol better captured what he saw as the most important feature of binge drinking—the desire to feel drunk. He thought that the nocturnal drinking model was problematic because (like many existing models) it created conditions under which researchers might mistake a mouse that was avoiding alcohol because it tasted bad for a mouse that was uninterested in becoming intoxicated. He argued that by altering the way in which DBA mice encountered alcohol, he could elicit a behavior similar to that in humans, where they would readily gulp down large enough quantities to get drunk.

Dr. White favored experimental scenarios where mice could freely choose between alcohol and water at all times and learn about the effects of drinking alcohol over a period of several weeks. Free choice and learning about alcohol's effects were both central enough to his understanding of why humans drank to excess that he was even willing to sacrifice high blood alcohols levels in his animals to capture these two aspects of the human situation. In an interview (ironically held in a bar), he told me about a study he had recently completed where he used a protocol combining aspects of the nocturnal drinking model with aspects of a two bottle choice experiment. He gave mice a bottle of alcohol for thirty minutes every day but without removing the water bottle so that they would have the choice of drinking water at all times. He thought that this study offered more convincing evidence that the mice were drinking because they wanted to experience the sensations of being drunk, but under these conditions the mice only drank to BACs of 0.06 to 0.07 gram percent—that amounted to the level of intoxication we might feel if we each finished our pints of beer in less than half an hour on an empty stomach, he told me. He admitted that this did make it difficult to get his papers published: Reviewers "hated" seeing those blood alcohol levels, he said, because, "You could still get in a car and drive, right? It's definitely not the kind of range that you'd see in alcoholics." While Dr. White's experience in peer review suggests that his position on what made a good model of binge drinking was not a widely shared one, it is notable that at least some alcohol researchers were willing to forgo the BAC link entirely and use other kinds of links to build up their epistemic scaffolds.

At first glance, it may seem that what underpins these dissenting statements and different experimental systems are different understandings of problem drinking. It could be, for example, that Drs. James and White believed the core feature of binge drinking was the intent to experience the intoxicating effects of alcohol, while the rest of the ARG saw the most important feature as the pharmacological dose of alcohol that a drinker received. Through talking with various ARG researchers about their reasons for including or excluding particular features of binge drinking, though, it became clear that their understandings of binge drinking did not differ greatly, but their epistemological commitments did. To some, the benefit of working with animal models was that they could separate out the various motivating forces that existed together in the human situation and then study them each in isolation. To others, animal models needed to include as many features of the human situation as possible in order to be useful. The key fault lines within the group, then, were not ontological commitments to different understandings of binge drinking but different opinions on how best to use mice to model human behavior.

Dr. James, for example, recognized that the taste of alcohol was an important factor in human drinking and readily acknowledged that eliminating that aspect of drinking meant he detracted from the "face validity" of his model. But he told me that he was willing to forgo some connections to the human scenario in order to produce new kinds of data:

> Maybe [the intragastric drinking model] doesn't have the same face validity as having a human drink to that point, but there's nothing that says once they're at that stage that the mechanisms that would be involved in having them maintain their intake wouldn't be the same as what you might expect in a human or other mammal. So basically we're pushing them to the point where—because up until now, there's really only been one mouse genotype that's worth studying, and that's the C57. And I already think that there's a problem with that. Although we've only looked at probably four or five different genotypes now, I think that what we're going to find with the intragastric that there are more interesting genotypes out there than C57 to pay attention to.

He admitted that his model may not be as useful for studying *how* humans develop drinking problems, but it could still be useful for understanding the biological changes that happened once they became heavy drinkers. Moreover, by making trade-offs around taste, he was able to produce scenarios that resembled human drinking in other important ways. He thought models using only C57 mice—which included the nocturnal drinking model—were extremely limiting because this approach did not allow researchers to capture information about genetic variation. Humans were hardly genetically identical like C57 mice, and yet researchers continued to use them because they were the only inbred strain that would drink readily. In his view, then, he was simply making a different trade-off than other researchers, trading some face validity in the initiation phase to get more information about how different genotypes responded to heavy drinking. In an ideal world, researchers might not have to make this choice, but the mouse's unique sensitivity to the taste of alcohol placed some limits on its capacities as a research tool.

Although he had chosen a very different experimental setup, Dr. White agreed with Dr. James on many of these points. Dr. White pointed out that his experimental design also made trade-offs around taste, although they were different ones than Dr. James made. Dr. White noted that while the taste of alcohol is an important part of the equation for humans, very few people start out drinking by sipping ethanol straight from the bottle. When humans first start to drink, he explained to me, we tend to choose sweet, flavored drinks that mask the taste of alcohol. By offering his mice cherry- and grape-flavored drinks lacking the alcoholic burn, Dr. James' setup was in some respects quite similar to some human binge drinking scenarios (as anyone who has had too much Purple Jesus at a college party can attest). Dr. White explained that although he recognized masking the taste of alcohol was an important component of human drinking, he chose to offer his mice unsweetened ethanol solution to avoid introducing a confounding factor into his experimental scenario. He wanted to make sure that his mice were drinking because they wanted to become intoxicated, and not because they liked the grape flavor or mistook the artificial sweetener for a source of calories. Where Dr. James and Dr. White parted ways, then, was not at their understandings of binge drinking but at their understandings of which aspects of taste—the taste of alcohol or the taste of flavored solutions—presented the most serious confounds for knowledge production.

The discussions that took place in the ARG more generally also revolved around methodological problems, not around the nature of binge drinking itself. To Dr. Linda Anderson, a behavioral pharmacologist who was new to the alcohol research field when she first joined the group, the terms of this debate seemed strange. She was an enthusiastic early adopter of the limited water model, and she was rather surprised when the ARG decided to abandon the limited water model, given the group's stated goal to develop a protocol where mice would reach intoxicating BACs. She summarized for me why she believed the group took issue with the limited water model in particular:

> It creates a motivational confound. Ideally, you want an animal model of alcoholism where the animal likes or wants to consume that alcohol not because they're thirsty but because there's some property of the alcohol that they enjoy. And so whenever you introduce food restrictions or fluid restrictions, or even sweeten the alcohol, you have these confounding motivators. You don't want the animal drinking solely because they're thirsty. You don't want them drinking because you've disrupted their caloric balance by taking away all of their food. You don't want them drinking just because they like the sugar. So that's why the [nocturnal] model is more appealing, because you don't have those confounding factors, and it's cleaner. They're drinking because they're drinking.

Dr. Anderson did not agree with this position. To her, fluid deprivation seemed to be a perfectly legitimate technique for getting mice to drink heavily, and the prohibition against it within the alcohol community seemed like an irrational taboo. She described the decision to stop developing the limited water model to me as a "scientific-political" decision: "[The ARG abandoned the model] pretty much because of the fluid deprivation issue, which is a big no-no in the alcohol field—although alcoholics are clearly fluid deprived, these people do not drink water and hydrate themselves properly—but nevertheless, alcohol folks have established over decades of research that fluid deprivation is bad. So that's when we met, and as a group, and it was decided the [limited water] model was no longer a high-priority model." In this quote, Dr. Anderson formulated her disagreement with the decision to abandon the limited water model as an ontological argument—the limited water model is valid because alcoholics are themselves fluid deprived. That argument, however, failed to address the epistemological terms on which the decision to abandon the model within the ARG was made. This debate shows why it is important to distinguish between reductionism as a statement about the world versus reductionism as methodological strategy. In a scenario where Dr. Anderson's colleagues believed that dehydration was not an important component of human drinking, engaging them in a discussion about how much water alcoholics drink might have had a chance of success. But in this case, her fellow researchers agreed that alcoholics are in reality often fluid deprived, but disagreed that fluid deprivation was a good means of producing data in a mouse model. The decision to narrow in on a few motivating factors was a pragmatic, methodological one, not an expression of a belief that human binge drinking could be explained by only these few factors.

To call the ARG's models "reductionist," then, is correct insofar as reductionism can be understood as an epistemological strategy, but it is perhaps easier to use a different term. In his study of global climate science, Paul Edwards (2010) has described the models his researchers use as "reproductionist" rather than "reductionist." Edwards (2010, 280–81) explains, "The reductionist ideals of an earlier age of science sought always to explain large-scale phenomena through smaller-scale component processes. Complementing rather than replacing it, in computational sciences such as meteorology a new ideal has emerged—an ideal we might call *reproductionism*. Reproductionism seeks to simulate a phenomenon, regardless of scale, using whatever combination of theory, data, and "semi-empirical" parameters may be required."

Reproductionism, in Edwards's (2010, 283) description, is a pragmatic approach that treats both hard data and approximations as valuable for meeting "here and now" knowledge needs, rather than aiming to generate a "single, fixed truth, valid for everyone, everywhere, at all times." Climate researchers assume that their data sets or models will never do full justice to global weather systems, and therefore the outputs of each individual model will be flawed in its own way. Edwards argues that this reproductionist ethos has led to a proliferation of models and alternative data images, where weather forecasters favor "ensemble forecasts" that combine the results of multiple different models rather than relying on a single, trusted model.

The ARG researchers' efforts are also, I argue, better understood as "reproductionist" than reductionist. Researchers saw themselves as attempting to simulate an aspect of a complex disorder without assuming that they could re-create all of its features. Because they understood the binge drinking model they were developing as partial, they tolerated and even welcomed other models that might capture other facets of that behavior.[9] Even in a venue such as the ARG where researchers were explicitly tasked with focusing their collective efforts on a single model, the tendency for models to proliferate rather than coalesce continued. It took pressure from reviewers and the budget constraints of the NIAAA to focus the group's efforts back on the nocturnal drinking model. This individual model may be thought of as reductionist because it ignored some aspects of the human condition and focused intensely on others, but no model on its own was meant to represent binge drinking in its entirety. Animal modeling was an ensemble project that researchers saw as stretching far beyond the boundaries of the ARG, a project where researchers hoped that generating multiple partial drinking phenotypes would lead to a better picture of the complex whole.

THE FORCES SHAPING MOUSE (AND HUMAN) BEHAVIOR

Examining the epistemic by-products researchers produced while developing the nocturnal drinking model further complicates the picture of how this model contributes to understandings of human drinking. As researchers built up the model, adjusted its parameters, and began to use it as the foundation for new research projects, they gathered an abundance of information about the forces shaping a mouse's consumption of alcohol—some biological and some not. The genetic, neurobiological, and pharmacological information they sought, along with the environmental knowledge they acquired along the way, also contributed to an understanding of binge drinking as a complex phenomenon.

Alcohol researchers' existing store of epistemic by-products was key to bringing the new model into being in the first place. As Dr. Wilson, one of the group's founders, put it to me, there was "nothing really superspecial" about the nocturnal drinking model. Researchers had long known that mice tended to consume most of their food and water a few hours after waking up, that offering alcohol to mice for a limited time and on a regular schedule would increase the amount of alcohol they drank, and that restricting their food or water would also increase their alcohol intake. What was new about the model, Dr. Wilson said, was the way that the group had combined and systematically manipulated the various parameters involved—the amount of time water was available, the precise time at which the water bottles were switched for alcohol bottles, the concentration of alcohol offered to the mice, the amount of time they were left on the cage—and characterized the behavior that resulted. Researchers reworked knowledge and techniques previously considered by-products into a valuable product by aggregating and formalizing that knowledge.

Knowledge about the effects of the laboratory environment on mouse behavior was especially important for the ARG's work because genetic factors alone did not seem to be enough to engender the type of drinking researchers sought. Published data on the amount of alcohol that various mouse strains would drink had been available since the late 1950s (McClearn and Rodgers 1959), but even the heaviest drinking strains, such as the C57s, seemed to modulate their drinking so they would not become intoxicated for extended periods of time (Dole and Gentry 1984). Only by placing genetically inclined strains of mice in particular experimental situations did they begin to drink past the 0.08 percent threshold and show behavioral signs of intoxication. In this sense, the new binge drinking models were not simply genetic or biological models for the human but "situated models," where animals could not be understood as models for human behavior in isolation from the environmental and experimental circumstances in which they were studied (Ankeny et al. 2014). As we have already seen, researchers put a great deal of thought into the problem of how to craft an experimental situation that would elicit the right kind of drinking behavior. It was only when researchers situated mice in particular experimental situations that could be understood as eliciting a desire to drink for "pharmacological effect" that they became credible models for human binge drinking.

As researchers worked with the models, they gained new epistemic byproducts that further contributed to their understanding of both the models and binge drinking. When I spoke with Dr. Anderson about her experience with the nocturnal and limited water models, for example, one of the things that she emphasized was their relative insensitivity to variation in the laboratory environment. She told me:

> That's the one thing that I have to say is that the beauty of these models, both of them, [limited] and [nocturnal], is you can take that model and apply it in any place, and they will at least drink 0.08. We're getting numbers that are bang on, we're getting bang on with what [other researchers in the ARG] have found. And we've got undergrads running the experiments, we've got grad students, we've got people who are probably hung over when they're doing the experiments, and they're getting the same type of data. So it is a beautifully robust model.

In contrast to many behavioral tests, where differences in the results often emerged when researchers performed the tests in a new laboratory, the new binge drinking models seemed to reliably induce mice to drink to high BACs in a variety of laboratories and with a variety of experimenters. To Dr. Anderson, this signaled that the models were tapping into a strong biological impulse. As she put it to me:

> Three hours into the dark cycle—that was not selected willy-nilly. There's studies on feeding and drinking behavior and the circadian rhythm, and that is the time when typically mice and rats consume most of their food and water. That's an innate cycle that they have for regulating their intake,

> so when you present booze at that time point, the animals already have an internal inclination to eat and drink at that time. So if that booze is the only thing that's available—that's what's going on with these models, right? One bottle, boom. They're going to drink, because that's what their internal clock is telling them to do.

Dr. Anderson's description of an "innate cycle" and an "internal clock" for regulating drinking places the forces driving this behavior squarely in the biology of the mice, in a place where external factors such as the person changing the bottles on the cages seemed to have little impact. But the robustness of the model comes only through the alignment of the scheduled drinking window with those biological rhythms, a joint accomplishment of both biology and the environment that results in drinking to excess.

Work with the models generated evidence of variation as well as consistency. Contrary to the claims of Rose and Abi-Rached (2013, 86), who express concern that the "differences among individual living organisms . . . disappear as the animal is transformed into a series of measurements," researchers preserved some of these differences in their publications. For example, as researchers worked with the nocturnal drinking protocol, they observed that some mice drank much more than the 0.08 gram percent threshold, reaching BACs up to 0.14 gram percent. In the initial publication describing the new protocol, the authors noted these individual differences that emerged even in groups of genetically identical animals, suggesting (in contrast to Dr. Anderson's claims) that something in the laboratory environment did indeed have an effect on the mouse's drinking behavior. They argued that this variation was consistent with the idea that even the genes influencing the C57's well-known propensity for alcohol drinking operated in a "permissive" rather than a deterministic manner. The design of the nocturnal protocol also gave researchers inadvertent opportunities to observe differences in the drinking habits of individual mice over time. The protocol called for the mice to be offered alcohol every evening for five evenings—a design choice that researchers made to acclimate the mice to the presence of the researchers and the alcohol-filled sipper tubes, so that the novelty of these factors would not impact how much they drank. This acclimation procedure, however, also produced data that allowed researchers to see how an individual mouse's drinking habits varied over time. ARG researchers noted that BAC levels among C57 mice in the nocturnal drinking protocol were not simply variable; some mice reliably drank more alcohol than others. Over the course of the five-night study, the same mice repeatedly reached high BACs, prompting ARG researchers to speculate among themselves about developmental or learning processes that might account for these enduring differences even among genetically identical individuals.

Generating data on individual variation was not what researchers considered their main line of knowledge production work. What the ARG members intended to do with their new model was produce biological information that would contribute to clinical treatment. One of the first projects the ARG undertook after establishing the nocturnal drinking model was to examine changes in gene expression in the brain that took place as mice repeatedly "binged" on alcohol. They compared the results of this study to results from other animal models and to genome-wide association studies (GWAS) in humans, looking for overlaps between genes whose expression changed in animal models and gene variants that were overrepresented in human alcoholics. Some of the genes that emerged from this analysis, such as a cluster of genes associated with inflammation, were new to alcohol researchers. The ARG created a "priority list" of these genes, which they hoped would act as a point of coordination between researchers using other model organisms and researchers doing clinical work. ARG researchers also began a selective breeding program, testing genetically heterogeneous mice in the nocturnal drinking test and mating the ones who drank to especially high BACs with each other. After several generations of breeding, they had produced a line of mice that would drink to 0.15 gram percent on average. The researchers planned to use this unique line of mice to identify more genes associated with heavy drinking, to understand which brain regions were necessary for heavy drinking to occur, and to investigate how early life stress might modulate drinking. Elsewhere within the ARG, another group of researchers started studying the mechanisms of action of the group's priority list of genes, looking at where they were expressed in the brain and how those brain regions interacted with circuits already known to impact drinking. More recently, the ARG researchers have begun to use the nocturnal drinking protocol as a drug screening tool. Members of the ARG with expertise in bioinformatics developed an algorithm for searching databases of existing Food and Drug Administration–approved drugs to identify those that might hold promise as treatments for alcohol abuse, and researchers with expertise in animal behavior planned to then administer those drugs to mice to see if any would reduce the amount of alcohol they consumed in the nocturnal drinking model.

These projects all fit neatly within in a biomedical research paradigm, and it is easy to see how they follow from and reinforce a conception of alcoholism as a genetic or brain-based disorder. Less obvious, though, is the way in which researchers' work with these models created opportunities for researchers to think about human drinking in nonbiological terms. The environmental knowledge that the ARG members drew on and produced as they crafted the nocturnal drinking scenario also contributed to an understanding of binge drinking as a situational phenomenon.

Consider, for example, the following excerpts from an interview with Dr. Wilson, where he offered two different explanations for why offering alcohol at mealtime might lead to heavy drinking. Dr. Wilson told me that he remained agnostic about why mice drank as much as they did in the nocturnal model, but he had a few hypotheses. One was that alcohol might reduce a drinker's appetite, making it easier for them to delay or skip a meal and keep drinking instead. Conversely, alcohol might act synergistically with food, with the consumption of one amplifying the desire for the other:

> Presumably, it's similar to one of us going to have a glass of wine before dinner but suddenly ending up with three, OK? So you're on an empty stomach, you may forget how many you've already had, or you like the feeling that you've gotten, or you've lost your appetite for the dinner because you're drinking, so you just keep drinking. I don't know why mice do it, too, but they do.
>
> It's kind of analogous to the bar situation, where they're delivering peanuts that are salted, and you drink a lot of beer while you're eating the peanuts. When you go home, you don't just keep drinking alcohol, unless you're an alcoholic. But in the bar you're choking [the drinks and the peanuts] down.

Though these explanations identified different factors that contributed to heavy drinking, Dr. Wilson did not represent human or animal drinking as a mere biological drive in either case. Rather, he described heavy drinking as a phenomenon that "require[d] certain contingencies to engage in," as he put it later on in our interview. Hunger around dinnertime or thirst from eating handfuls of salty nuts might both contribute to heavy drinking, but neither was determinative. At home, in the absence of salty snacks, the desire to drink diminished. If therapeutic interventions to help addicts manage their impulses can be thought of as "civilizing technologies" designed to rein in deviant behavior, as Vrecko (2010b) argues, then the carefully crafted experimental environment could be thought of as an "uncivilizing technology." In researchers' descriptions, it was the setting of the bar or the timing of the before-dinner drink that created a set of circumstances under which both mice and humans were apt to lose control of their drinking. This is a vision of drinking behavior that came not from genetic knowledge alone but from researchers' intimate familiarity with designing experimental environments to promote alcohol consumption.

Dr. Anderson similarly talked about the limited water model in a way that emphasized the structural features of drinking, both in animals and in humans. She told me that when she taught her undergraduate classes, she referred to the model as the "late for happy hour" model to help her students understand how it works: "I mean, who knows how much [the mice] are learning that the alcohol is only available for half an hour? [In my classes], I call it the "late for happy hour" model. You know, when you're late for happy hour, you order a whole bunch of drinks and you pound them all down, even though once you pay for them, they're not going to take them away." She hypothesized that mice in the limited water model drink so much because they might be learning over the course of the experiment that the alcohol bottle will be available for only a short period of time. That condition of scarcity was what appears to drive both human and animal drinking in Dr. Anderson's description. Indeed, Dr. Anderson presented a vision of a drinking human who seemed even more controlled by her environment than the drinking mouse: while at least a mouse in a limited water experiment would be correct in surmising that it has to drink quickly or the alcohol bottle will disappear, the human presumably knows that her drinks will not be taken away once she has paid for them but drinks them in rapid succession anyway. Notably, Dr. Anderson used description details of the limited water protocol itself—not the data the model produced—to paint a picture of the forces motivating mouse (and human) drinking.

CONCLUSION

In this chapter, my aim has been to show how embedding the nocturnal drinking model in the context of the ARG's efforts to develop and validate it makes it easier to understand how researchers could see themselves as unraveling the complexity of behavior while studying individual mice trapped in cages with alcohol bottles. The extensive discussions within the ARG about motivation show that much more was at stake than the satisfaction of a simple blood alcohol biomarker. Many researchers wanted to capture other facets of human binge drinking in addition to BAC, and those who were willing to forgo particular resemblances did so for epistemological reasons rather than out of a belief that human binge drinking was a simple matter of pharmacological exposure. Researchers' experiences of seeing stability and variation in their animals' drinking patterns also reinforced and challenged the idea that heavy drinking is a behavior under genetic control. The ARG researchers knew well that genetic predisposition alone was not enough to make most mice drink heavily—recognition of the mouse's intrinsic reluctance to drink was what started the entire project in the first place. By combining genetic susceptibilities and environmental manipulations in order to get mice to drink heavily, researchers believed that they had created a model that reproduced some of the biological and structural factors that led to humans to binge drink as well. Considering researchers' epistemological commitments, scaffold work, and epistemic by-products together with the nocturnal drinking model itself helps us understand how individuals could reach very different conclusions on what this work has to say about human behavior.

Two contrasting examples illustrate the extent of this interpretive flexibility: in a recent article examining knockout mouse research on maternal behavior, Philip Rosoff (2010, 202) objects strongly to these experiments, which he sees as grounded in the hypothesis that genes are causally related to particular behaviors. He aims to debunk this hypothesis in his article by "laying bare [the] history and complexity" of these experiments, thereby demonstrating "the emptiness of any scientific or lay claims to strong deterministic claims" (204). Gail Davies (2010, 68), meanwhile, examines a very similar set of knockout mouse experiments on fear behavior but reaches the conclusion that "there is growing awareness of . . . complexity and multiplicity in much laboratory science." In light of these observations, she concludes elsewhere that critical commentaries on genetic reductionism and determinism now seem "somewhat beside the point," in that they are "based on simplistic assumptions about genes that have been superseded. We now find ourselves somewhere different" (Davies 2013, 276).

The interpretive flexibility of these kinds of experiments—which could be read as either examples of egregious genetic determinism or a movement in the direction of complexity and multiplicity—explains why some critiques of addiction genetics may be ineffective in engaging scientists. Critiques grounded in assertions of complexity may appear to scientists not to be critiques at all but merely accurate statements about the difficulties they as practitioners encounter in trying to study behavior. A recent exchange in *The Lancet Psychiatry* demonstrates this point. In a personal viewpoint article, several researchers put forward a critique of the brain disease model of addiction, pointing to scientific evidence that they believed contradicted the hypothesis that addiction is a chronic relapsing disorder. They pointed out, for example, that "popular accounts of [animal] studies underplay the extent to which the results depend on specifically bred strains of rats and the conditions in which the animals are housed" (Hall, Carter, and Forlini 2015, 106). They argued that this evidence suggested that "addictive behavioural patterns are not invariably the outcome of chronic self-administration of drugs in animals" (106). But while the authors saw this evidence as something that weakened the brain disease model of addiction, this was not how NIDA and NIAAA directors Nora Volkow and George Koob, respectively, took it. In a reply to Wayne Hall and colleagues, they agreed that not all animals or humans would develop addictions in all circumstances. "However," they continued, "we do not understand why this fact should negate the value of the disease model in addiction" (Volkow and Koob 2015, 678). They saw this heterogeneity as perfectly compatible with their understanding of what it meant to call addiction a disease and argued that animal models were valuable precisely because they could shed light on who was most at risk of developing an addiction and how addictions transformed from mild to severe forms. Rose and Abi-Rached's (2013) more general suggestions for how psychiatric animal models could be improved face a similar problem. Rose and Abi-Rached propose a more expansive approach to animal modeling, one where "the capacities, pathologies, and behaviors of animals, like those of humans, would have to be located in their form of life and their constant dynamic interchanges with the specificities of their milieu" (104). While they clearly intend this statement to be a call for a different type of modeling work, an animal behavior geneticist might read this it as simply describing the work that they already do. The nocturnal drinking model, after all, attempts to take the unique sensitivity of rodents to the taste of alcohol into account and to modify it through ongoing exposure to a new environment. The interpretive flexibility of genetic animal models means that the kinds of statements I have pointed to here could be seen as either calls to entirely reimagine the field's practices or accurate descriptions of research as it exists today, depending on the vantage point.

This conundrum is especially apparent in commentaries that draw on biology in order to critique biology. Analysts approaching animal behavior genetics in this way seem to struggle with the question of how scientific practitioners can produce research that analysts so strongly disagree with, while at the same time producing other statements and findings that align with analysts' critical views. Rose and Abi-Rached (2013), for example, take issue with one practitioner's "breathless translation between species," while simultaneously observing in a footnote that this same practitioner has "elsewhere given some serious and thoughtful consideration" to problems in animal modeling.

And while they frame their chapter with the provocative assertion that there is something "wrong" with behavioral animal research, they note that many of the concerns they raise have already been articulated by "careful" researchers in the field (108). Rosoff (2010) also struggles with this tension. In some places he lambasts the field of behavior genetics and its practitioners, arguing that "by geneticizing phenotypes which may be immune to such causal attribution, they pervert their methodology in the vain pursuit of a concrete, discrete and objective answer to an inherently diffuse, massively multifactorial and subjective question" (228). But in other places, he notes that practitioners are often quite circumspect in formulating their claims, and instead blames the problematic notions he wishes to critique on "less well-informed readers" (210) or "secondary interpreters" such as the popular press (214). Evelyn Fox Keller (2010) points backward to Francis Galton for the origins of the problematic nature/nurture thinking she aims to eradicate, but then she looks forward to "the new science of genetics coming out of today's research laboratories" for a "route out" (73). These attempts to use science to testify against itself strike me as an ineffective way of engaging with scientific practitioners, for whom translations between species and methodological cautions are not contradictory but part of a coherent whole. Even when arguments of this kind are aimed at other audiences, they face a significant barrier to success—they require nonscientists to establish themselves as more credible interpreters of scientific evidence than scientists, which is a high bar to clear in a culture where scientists enjoy substantial authority as arbitrators of meaning of scientific evidence (Hilgartner 1990).

I am not the first to note something like this problem. In his discussion of the "narrative of enlightened geneticization" that he identifies in schizophrenia research, Hedgecoe (2001) makes a similar argument about the ineffectuality of determinist and reductionist critiques. He writes, "Simply to criticize modern molecular genetic research in schizophrenia as 'deterministic' is to ignore the role that it claims to offer to non-genetic factors. To suggest that there is no single gene 'for' schizophrenia in the belief that this undermines molecular approaches is to miss the point that genetic researchers are the first to admit this" (Hedgecoe 2001, 902). Sara Shostak (2013, 18) has similarly argued that it is now time for social studies of genetics "move beyond the geneticization thesis," lest analysts miss "the opportunity to observe profound changes in how genes, environments, and human bodies are conceptualized and operationalized in human research" (19).

My claim is a bit different. I am not arguing that critiques of geneticization or medicalization have wholesale "run out of steam," as Latour (2004) has provocatively put it, but that they are less effective insofar as they rely on ontological claims about the complexity of the world and the importance of nongenetic factors. Ontological assertions are ineffective because they too closely resemble many scientists' own beliefs about the natural world. But just as scientists' discussions of complexity can be understood as expressing both ontological and epistemological commitments, so too can analysts' critiques. The geneticization thesis as originally articulated by Lippmann (1992) is not just an argument for seeing health as a historically contingent, socially shaped phenomenon; it is also one about the efficacy of biomedical interventions and the allocation of scarce resources. Hall, Carter, and Forlini (2015) similarly express concern that the brain disease model of addiction might lead to a neglect of health policy research, in addition to their criticisms of the brain disease model itself. They argue that even if biomedical research does produce new therapies, this would not negate the need for preventative public health measures or for psychosocial support systems (see also Carter, Capps, and Hall 2012). These arguments about the allocation of resources and proper mix of treatment strategies are not ones that depend strongly on a specific view about the underlying reality of addiction; they are arguments about how best to research it and intervene on it. These arguments can be made without engaging in debates about whether scientific practitioners *really* believe in the complexity of addiction, and might be more effective in reaching scientists if framed in purely epistemological or pragmatic terms. The ARG researchers' conception of their mission—to contribute one model to an ensemble effort to understand binge drinking—is one that offers openings for engaging in methodological debate. Taking the ARG researchers' beliefs about complexity at face value, scholars could still argue about how to achieve the right mix of biomedical research, social science research, and harm reduction approaches to binge drinking within this ensemble effort. At the very least, criticisms along these lines could not be dismissed out of hand by practitioners as simply misunderstanding the science.

The argument I have advanced here about the interpretive flexibility of animal modeling work also adds new weight to some long-standing criticisms of addiction research. In critiquing the disease model of alcoholism, analysts have first and foremost been concerned with the impact of biomedical research in nonscientific venues—in popular culture, public policy, or drinkers' own perceptions of their behavior. The source of the interpretive flexibility I have identified here suggests that it is possible, and perhaps even likely, that researchers who care a great deal about the complexity of addiction will still end up contributing to simplistic understandings of human behavior.

Campbell (2010) makes this point in her insightful analysis of movements toward a "critical neuroscience." She argues that contemporary notions of addiction as a form of "disrupted volition" and neuroimaging tools have the transformative power to change stigmatizing cultural views of addiction, but that these same notions "could as easily be cast in into a static register of biogenetic determinism" (100). Campbell calls on critical neuroscientists to find ways of "taking into account the cultural meanings of its concepts and representations" (100), but how exactly they might do this is unclear. Examining the mechanisms through which interpretive flexibility is generated inside the laboratory offers useful insights for those who might want to intervene in this process. For example, my ethnographic analysis suggests is that the extrafactual work of animal modeling is an important part of what gives those models their interpretive flexibility; therefore, those who are not engaged in modeling work might take away very different understandings of behavior from those models than the practitioners who work with them. Clinicians, policy makers, or members of the general public are highly unlikely to have access to the full range of discussions about the validity of particular mouse models, or to any of the environmental knowledge that researchers gain through their work in the laboratory. For those who are not working in the laboratory, animal models are a different kind of object with different kinds of representational affordances. It is to the differences between the conclusions that animal behavior geneticists draw based on modeling work and the way that their research is represented in other knowledge communities that we turn to next.

CHAPTER 6

Leaving the Laboratory

Near the end of my stay at Coast, Dr. Smith invited me to help him lead the session on "ethics" as part of the introductory behavior genetics course. One of the readings I suggested came from *The DNA Mystique*, Dorothy Nelkin and Susan Lindee's (1995) widely cited book examining popular representations of genetics. Nelkin and Lindee catalogued the resurgence of "gene talk" in the late twentieth century and analyzed how these genetically determinist and reductionist stories reaffirmed existing social prejudices about antisocial behavior, gender, and racial difference. Because popular references to behavior genetics research figure prominently in the narrative, I was worried that Dr. Smith and the graduate students might see Nelkin and Lindee's argument as an accusation that behavior geneticists themselves held deterministic views or were contributing to discriminatory attitudes and social policies. But to my surprise, they loved the selection. They read Nelkin and Lindee's analysis as evidence that the public is determinist or reductionist in how they think about genetics and misinterprets behavior genetics research accordingly. As Dr. Smith joked to the class, the public seems to be "hard wired to believe in genes." Dr. Smith seemed to enjoy the selection enough that I bought him a copy of the book as a thank-you gift after I had left Coast.

Behavior genetics research has generated much critical commentary because of its perceived potential to negatively impact public policy and perceptions of vulnerable social groups.[1] News media coverage of this research has received particular scrutiny. A recurring theme from analyses of articles covering behavior genetics is that they overemphasize the centrality of genes and downplay uncertainties and nongenetic factors. In their study of media coverage on the genetics of alcoholism, for example, Peter Conrad and Dana Weinberg (1996) made the tongue-in-cheek observation that the "gene for alcoholism" was discovered three times between 1980 and 1995. They showed how news reports announced the discovery of such a gene at three widely separated time points and argued that in each case, these articles overemphasized the contribution of the genes under investigation and presented undue optimism about the possibility for new treatments. In researching media coverage of the genetics of other psychiatric disorders, Conrad (2001) found that this pattern was relatively consistent. News reports used a frame that he termed *genetic optimism*, which emphasizes that genes for particular disorders exist, can be found, and that finding them will have positive outcomes. Conrad (2001) argued that this frame showed signs of weakening at the turn of the century, but other research suggests that it remains prevalent. Molly Dingel and colleagues (2014) argue that news media articles on the genetics of addiction have continued to foreground genetic factors and offer only cursory treatments of environmental contributions. Even when discussing studies explicitly designed to take both environmental and genetic factors into account, analysts argue that news media accounts still depict behavioral disorders as genetic (Horwitz 2005).

Behavior geneticists largely agree that news media coverage of their research is problematic. They also see popular portrayals of their work as overly reductionist and deterministic and have voiced concerns about the social consequences of these statements. In a *Science* magazine opinion article, for example, several prominent behavior geneticists took issue with the "gene for" framing that often appears in news reports (McGuffin, Riley, and Plomin 2001). Much like critical commentators from outside the field, they argued that this phrase misleadingly implies that there is a direct relationship between particular mutations and traits such as homosexuality, aggression, or criminality. Researchers at Coast were very much aligned with this view. As practitioners who believed that behaviors were fundamentally complex and that simplistic claims could be harmful, they were distressed by the ubiquity of the "gene for" trope in news media articles covering their research.

Classic science and technology studies (STS) accounts of science communication have emphasized the agency of scientists in authorizing journalistic accounts of their research, and yet they have also noted scientists' frustrations with this process. While scientists may hold a privileged position that allows them to adjudicate between "appropriate simplifications" and "distortions" of their work (Hilgartner 1990), they nevertheless feel as though they have little power when sitting face to face with a journalist or reading a newspaper article describing their research. This chapter aims to make this paradox more comprehensible by contrasting aspects of the research culture at Coast that I have described so far with the conventions of journalism. I focus my analysis at the level of individual scientific actors rather than institutional, political, or economic concerns shaping media coverage, examining how researchers at Coast participated as "performers" in public venues and "commentators" on media coverage of their research (Hilgartner 2011). In addition to ethnographic material, I draw on a set of articles published in major North American newspapers that reported on animal behavior genetics research (see the appendix for more information on the construction of this data set). I argue that while scientists as a social group may enjoy a strong degree of control over how their research is presented in the media, individual researchers nevertheless feel as though they have little control because their favored techniques for managing their claims are ineffective in these venues. The cautious language that researchers at Coast used to signal their epistemic commitments, for example, translated poorly in a journalistic context, where it seemed like typical scientific jargon. Likewise, the methodological work they did on the scaffolds of their models and the epistemic by-products they generated were difficult to circulate through news reports, especially in comparison to genetic findings. Practitioners engaged in various forms of "repair work" to refine and restrict the role attributed to genes in news media articles, but engaging in this work only reinforced their perception that they had little power to shape popular representations of their research.

CONTROLLING ANIMAL BEHAVIOR GENETICS NARRATIVES

> The feel-good gene: Those of us who don't have this natural bliss benefit are more likely to be anxious, and to self-medicate.
>
> *New York Times* (FRIEDMAN 2015)

> Alcoholism could be in our DNA, experts have suggested, after a gene linked to excessive drinking was discovered by scientists. A single mutation in the gene can scramble the chemical messages which inhibit drinking, compromising the body's ability to consume alcohol in moderation.
>
> *National Post* (COLLINS 2013)

> Scientists may have found a way to make you forget you're addicted to meth. . . . Researchers in Florida have discovered a method of wiping away memories, using a specific chemical.
>
> *Washington Post* (MILLER 2015)

These three news media headlines and ledes are just a sampling of the kind of news coverage of addiction research that "can really make behavior geneticists crazy," as Dr. Smith put it to me. One way that actors and analysts alike have responded to such statements is to ask questions about the source(s) of these reductionist and determinist ideas. Scientists have tended to place responsibility for problematic statements with journalists or the lay public. McGuffin and colleagues (2001, 1232), for example, write, "Rarely is it mentioned that traits involving behavior are likely to have a more complex genetic basis. This is probably because most journalists—in common with most educated laypeople (and some biologists)—tend to have a straightforward, single-gene view of genetics. But single genes do not determine most human behaviors." STS and science communication scholars are largely in agreement that "gene for" statements are abundant, inaccurate, and potentially harmful, but they have questioned scientists' claims about the origins of these statements. Researchers have challenged, for example, the idea that members of the lay public hold reductionist views of genetics. Martin Richards (2006) asked nonscientists to rank particular disorders on a scale from "totally inherited" to "totally environmental" and found that their answers for behavioral traits fell somewhere in the middle of the scale—much as a behavior geneticist's might. Respondents gave intelligence an average score of 2.4 out of 5, and rated antisocial behavior as predominantly environmental, with a score of 3.7.

Analysts have also challenged claims that journalists are responsible for exaggerating the significance or certainty of genetic findings. Scholars' arguments for the evolution of a "shared culture" (Dunwoody 1999) between scientists and journalists suggests that, if anything, journalists are even more likely to hold views on genetics that resemble those of scientists.[2] Tania Bubela and Timothy Caulfield (2004) examined news articles reporting on genetic discoveries published between 1995 and 2001 and found that the majority of the claims appearing in those articles were accurate reflections of statements made in scientific publications. In their judgment, only 11 percent of the news articles they examined made claims that were highly exaggerated compared to those made in the original scientific publications. Holly Stocking (1999) offers an excellent summary of the arguments and evidence for and against the more general contention that journalists tend to exaggerate the certainty of scientific findings. While some research has found that journalists tend to make science seem more certain than it is by eliminating caveats and historical context, other studies suggest that news media reports may also do the opposite—for example, reports may create the illusion of disagreement by giving equal weight to majority and fringe scientists. In turn, analysts have directed some of the responsibility for sensational claims about genes back at scientists themselves. Panofsky (2014), for example, has argued that some behavior geneticists have made a name for themselves by deliberately making provocative public statements about heredity and behavior.

Reductionist and determinist ideas about genes, then, seem to come from everywhere and nowhere. Arguments that attempt to locate the source of these ideas in particular social groups—scientists, journalists, or the lay public—largely fall apart on close examination. One behavior geneticist jokingly suggested that the ideas themselves seem to be imbued with agency, describing them as undead entities that continue to "rise from the grave" (Lerner 2006, 336). These attempts to analyze the persistence of reductionist thought or the degree of certainty of a scientific statement are problematic insofar that they treat these features as inherent properties of statements themselves rather than situational judgments. Just as Crist (1999) argues that anthropomorphism is a term that actors use flexibly to mark out what they consider inappropriate speculation about the animal mind, assessments of genetic reductionism or genetic "hype" are situational. Those involved in discussions about popular representations of behavior genetics use these phrases—typically disparagingly—to do work for their respective fields and positions. As I explored in chapter 2, researchers at Coast described their own work as grounded in complexity as a way of marking out differences between themselves and molecular biologists, but others disputed these categorizations. After leaving Coast, I visited another research group who dismissed the work taking place at Coast as "mere pharmacology" and described their own methods as the ones that truly embraced complexity. Both actors and analysts may claim to be able to objectively evaluate the degree of reductionism or determinism inherent in others' statements or methods, but it is only through successfully making such claims that they establish the supposedly independent yardsticks by which they make their determinations.

Stephen Hilgartner (1990) makes this point clearly in his widely cited essay on the "dominant view" of the popularization of science. He argues that in this view, scientists are responsible for creating knowledge and popularizers are responsible for disseminating "appropriately simplified" accounts of that knowledge. What counts as appropriate simplification versus "distortion," however, is not always clear; observers might make different judgments depending on their social location, interests, or their appraisal of the circumstances. Moreover, Hilgartner argues that scientific experts retain the authority to judge which popular accounts are appropriately simplified, which means that nonexperts "remain forever vulnerable to having their understandings and representations of science derided as 'popularized' and 'distorted'—even if they accurately repeat statements made to them by scientists" (Hilgartner 1990, 534).

The way that the Smith Laboratory reacted to Craig Venter's (2007) autobiographical book, *A Life Decoded*, illustrates Hilgartner's argument that scientists might interpret the very same statement as accurate or misleading, complex or reductionist, depending on the vantage point. Several members of the lab read the book while I was in residence at Coast and had mixed reactions to Venter's discussion of his family history of alcoholism. Venter (2007, 31) writes, "I do enjoy a drink now and then, even though there is a history of alcohol abuse in my family. The complications of alcoholism claimed the life of my grandfather at age sixty-three. His father died while drunk, run over while racing a horse and buggy. Could the susceptibility lie in our dopamine genes? Could my destiny have been shaped by a genetic repetition? In fact, I have four copies of the repeated section of DRD4, which is about average. Other genes are linked with dopamine, so DRD4 does not give the whole picture." How members of the Smith Laboratory judged this portrayal of genetic susceptibility to alcoholism depended on what message they thought was most important to convey to the public: that alcoholism was a complex disorder or that alcoholism had a biological basis. Dr. Smith was dissatisfied with this passage because he thought that it did not adequately describe the genetic complexity of addiction. Alcohol drinking does impact dopamine regulation, he told me, but it is only one of many brain systems that might be altered in addiction disorders. Dr. Lam was more sympathetic to Venter's description, because he believed Venter was overstating the case for DRD4's role in risk for alcoholism so Venter could emphasize that substance abuse disorders are not simply a lifestyle choice.

Hilgartner's (1990) account highlights the authority that scientists' evaluations of popular representations of their research carry in the public sphere. As the Smith Laboratory's discussion of Venter's book showed, researchers at Coast felt free to critique statements made by even very famous fellow scientists, to say nothing of the statements made by journalists. Other analysts have similarly emphasized the agency of scientists in shaping and evaluating mass media coverage of their research. Dorothy Nelkin (1987) has enumerated some of the ways that scientists exercised control over science news coverage in the 1970s and 1980s; for example, by refusing access to unpublished data or prepackaging information to encourage journalists to adopt their mind-set. In this way, industry scientists were able to shift the frame of media coverage of the ozone controversy from one of calamity to technical uncertainty. Academic scientists, she argues, were similarly successful in propagating stories that suited their institutional objectives and quashing those that they felt might undermine public support for federally funded research. Susan Lederer (1992) has described how animal researchers in the middle decades of the twentieth century similarly tailored the presentation of their research to diffuse critiques from anti-vivisectionists. She argues that researchers enhanced their apparent objectivity and downplayed the affective dimensions of animal research through the selective substitution of impersonal medical phrases for more evocative ones. Sharon Dunwoody (1999, 60) argues that overall, "scientific culture often succeeds spectacularly at determining the meaning of the science covered."

However, analysts have also noted that their descriptions of scientists' agency with respect to popular communication are at odds with scientists' subjective perceptions of their control over science news. By and large, scientists are suspicious or fearful of journalists and do not feel as though they enjoy great authority in shaping public discussions about their research. While researchers in many scientific fields express concern about how their work is portrayed in the popular press, behavior geneticists believed that their subject matter was especially susceptible to misinterpretation and misrepresentation. As McGuffin, Riley, and Plomin (2001, 1232) argued in their *Science* magazine opinion piece, "the genetics of behavior offers more opportunity for media sensationalism than any other branch of current science." Acknowledging this paradox, Nelkin (1987, 175) asked, "If reporting of science and technology is so uncritical, why is there continued tension between scientists and the press?"

I argue that one of the reasons that researchers at Coast felt as though they had little power in shaping popular messages about genes and behavior was that their favored techniques for controlling narratives within the scientific community were unsuccessful in public settings. Extending the time line of their research agenda, using cautious language, and preserving epistemic byproducts were all effective means of shaping fellow scientists' perceptions of the power of genes and the certainty of animal behavior genetics findings. In interviews with journalists or conversations with friends, however, these reliable strategies seemed to lose their power. Even if researchers had other powerful means of controlling news media messages at their disposal, such as the public relations office at Coast, the ineffectuality of their favored techniques for crafting careful claims made researchers feel as though they had little control over "gene for" messages.

CRAFTING CLAIMS FOR PUBLIC VENUES

From the perspective of researchers at Coast, there were three things that made it difficult to talk about animal behavior genetics research to the lay public: the threat of becoming the target of animal rights activists; the stigma surrounding psychiatric disorders; and the association of genetic research with eugenics, scientific racism, and exploitative practices. As one researcher joked to me, what made it difficult to talk about animal behavior genetics in public was the animal, the behavior, and the genetics.

As I argued in chapter 2, researchers at Coast used cautious claiming to navigate some of these difficulties. They sought to enhance the long-term credibility of their field by conservatively formulating the claims they made about influence of specific genes, the capacities of animal behavior genetics methods, the certainty of particular findings, or the distance to clinical applications. Cautious claiming was a strategy that worked well within the scientific community, especially when talking to researchers from neighboring scientific fields who held different views on gene action. But when talking about their work in public settings—such as in classrooms, with reporters, or with family and friends—researchers also saw compelling reasons to make stronger claims. Unlike fellow scientists, members of the general public did not necessarily share their assumptions that heavy drinking was an illness or that animal experiments could produce useful knowledge. Researchers at Coast thus felt torn between conflicting imperatives when trying to craft a claim for public consumption: they believed that overly broad claims were scientifically and socially dangerous, but they also believed that strong statements were needed to counter the messages of first, animal rights activists and second, popular beliefs that disorders such as alcoholism were a matter of personal weakness or poor life choices.

Animal rights activism was an important part of life at Coast University. As a research university in a liberal city in the United States, Coast was not infrequently the target of animal rights protests. Twice in my time at Coast, activists appeared at the entrance to the campus, and these protests were much discussed in the lunchroom for the days before and after.[3] Many students also rotated through a nearby nonhuman primate research facility in their first few years of graduate school, and they became acutely aware of the potential for public controversy to shape scientists' professional and personal lives through talking with researchers there. One senior researcher at the Primate Center who had mentored several Coast graduate students described to me how threats from animal rights activists had dramatically impacted the way she moved through the world: She changed her license plates frequently, wore a "disguise" with a wig and heavy glasses when she had her photo taken for her university's website, and her daughter was even placed under FBI protection for a period of time.

Mouse researchers were rarely the target of animal rights activists, but they felt a sense of solidarity with their besieged nonhuman primate colleagues. This solidarity was reinforced by the responses they received when talking about their research with friends, family, or members of the general public. During one of the lunchtime conversations about animal rights activism occasioned by a local protest, graduate student Kendra told me how she had made the mistake of mentioning that she worked at Coast when she was trying to adopt a cat, and the shelter employee made her sign a form saying that she would not use the cat for research. Kendra was deeply offended by that suggestion and also surprised that someone could so misunderstand the reality of laboratory research that they would think that she would go through a fairly lengthy adoption process to get an animal for research. In another instance, I was out socially with Madeline, one of the Smith Laboratory technicians, and I watched her nonscientist friends grow increasingly uncomfortable and quiet as Madeline told a story that involved her performing surgery on a mouse. Madeline was proud that she had been entrusted with a protocol that included a tricky surgery and seemed to realize only belatedly that her friends were too distracted by the surgery itself to share in her excitement about her career advancement.

The specter of animal rights activism promoted what Panofsky (2014) has described as a "bunker mentality" within the Coast community, where insiders were discouraged from making statements that resembled those of outside critics. While criticisms of animal behavior genetics methodology were entertained and even encouraged, criticisms about the use of animals in science were not. The laboratory space was full of signs, both subtle and overt, that animal rights sentiments were not welcomed. Just outside of the security door guarding the entrance to the laboratory was a large poster with a picture of a white mouse and the caption "They've Saved More Lives Than 911."[4] On my initial visit to Coast, Dr. Jackson took me on a tour of the building. As we reached the door of a room housing some especially valuable mice, she paused and asked me, "You're not a member of PETA [People for the Ethical Treatment of Animals], are you?" Fortunately for me, I misunderstood her question and thought she was inexplicably talking about *pita* bread (which we had a laugh over), but she was not the only one who suspected that I might be an undercover operative. In another instance, my vegetarian eating habits began to raise suspicion. Several members of the Smith Laboratory were part of a competitive barbecuing team and had decorated their workstations with images of cuts of meat. Sensing my lack of enthusiasm for their hobby and for meat in general, they began joking that perhaps I was an activist infiltrator. Worried about how these jokes might impact my nascent relationships with researchers at Coast, I took what was for me the rather drastic step of joining the lab for one of their weekly outings to a restaurant that had a bison burger special. After making a show of ordering and eating a burger along with the rest of the group, the teasing subsided.

Outside of the laboratory, researchers felt a responsibility to share the burden of publicly defending animal research that disproportionately fell to their nonhuman primate colleagues. Dr. Smith explained, "We're in kind of a long-scale war with PETA and the rest of the animal rights movement over the rationale and the validity of animal models at all. And you know, I could easily hide out here because nobody cares about rats and mice. I'm not using kittens or monkeys or dogs, something cute that people really get upset about. But it's really important to counteract that, and scientists more and more realize that we're doing the whole field a disservice by keeping our heads down on that score." Countering the arguments of animal rights activists, he told me, was one of the main reasons he put out press releases and gave interviews about his research. He felt he was established enough in his career that he did not need to promote his research through the media, yet he nevertheless believed it was important to provide examples of the benefits of animal research.

Practitioners at Coast felt a similar obligation to talk about their research in public settings because they believed nonscientists held problematic attitudes about psychiatric disorders. Researchers saw alcoholism and anxiety as biomedical disorders that were triggered by factors largely outside of individual control. But in their experience, their family members, friends, and students often believed that these behaviors were the result of poor decision-making or coping skills. Dr. Anderson, a researcher with the Alcohol Research Group (ARG), told me about a student of hers who wanted to do a senior thesis on a neuropsychiatric disease. Dr. Anderson suggested to the student that she first do some reading on various neuropsychiatric diseases and choose the one she was most interested in. Several days later, the student wrote to Dr. Anderson to say that she previously had no idea that addiction was considered a neuropsychiatric disease. The student said she had always thought that addiction was a "disorder of choice." Dr. Anderson continued, "And then she says [that she] worked at a rehab clinic! And I'm like, 'And it never occurred to you that these people are diseased? I'm sure you've got patients in your clinic that really, really want to be clean, and it probably isn't their first time in the rehab clinic, you know?' And I'm like, 'It never occurred to you that there might be a neurobiological problem?' So yeah, no, they're not aware." In Dr. Anderson's experience, the majority of her students did not start out with an "appreciation" of the role of biology in behavioral disorders. And when even firsthand experience with addicts did not unseat stigmatizing assumptions about psychiatric disorders, Dr. Anderson felt it was her responsibility as an educator to emphasize the importance of pathological biological processes. She told me that she often presents on her own work with animal models in her classes to try to change her students' views of psychiatric disorders and, as she put it, "prove there's genes that are regulating this behavior."

Researchers at Coast also equated the promotion of biomedical understandings of psychiatric disorders with the reduction of stigma. They saw public conversations about their research as a means of lessening the burden of disease sufferers by normalizing mental illness. Graduate student Emily told me that when she was considering taking a selective serotonin reuptake inhibitor (SSRI) while dealing with the breakup of a long-term relationship, she realized that she herself had internalized problematic cultural attitudes about mental illness. She found herself questioning whether she really needed help and why she could not just deal with the situation on her own. Emily told me that she realized that if she herself was unwilling to take SSRIs, then it made no sense to devote her life to developing biomedical therapies for psychiatric diseases. So, she decided both to take the SSRI and to talk more publicly about her experience and her research to help others.

However, using conversations about her research to transform nonscientists' views of psychiatric diseases was a tricky prospect. While researchers at Coast found it problematic when members of the lay public discounted the idea that behaviors had a biological basis, they were equally concerned about genetically reductionist or determinist views. Emily described to me the difficulty in trying to strike the right balance in talking about genetic contributions to behavioral disorders:

> I went on a date like maybe three weeks ago, and I was telling the guy what I did, and he was like, "Oh no, alcoholism isn't genetic." And I was like, "But—it is, because I study that! Like for a living." And he's like, "No, I don't believe that." And I was like, "Well—OK." And I do get a lot of that. . . . But when I try to explain, they'll be like, "So have you found a

> gene?" Either they don't really believe me or it's such simplified thinking that [they think] we're looking for one gene or something, and it's really kind of tough to adequately—adequately explain the complexity of the problem, I guess.

Graduate student Hannah described a situation that was the inverse of Emily's mansplaining date, where she felt her family members were overemphasizing the importance of heredity in their discussions about a cousin's risk for developing an addiction. In that case, she felt it was her role to emphasize the importance of environmental factors:

> I have a cousin who was adopted and has a history of drug abuse from his birth parents and all that kind of stuff, and it's like, well, he's clearly going to be susceptible, he's clearly at risk. And it's like, yes, but you also don't want to essentially jail someone, lock someone up to try to prevent the inevitable, because you don't know that it's inevitable. So I think people really understand the genetics, but I think they take a very simplistic view of it and don't really realize how flexible, how individual genes have very small effects on it. It's not, you know, you have this gene, you've got it. Upbringing can have a significant impact on that.

Like the molecular biologists of animal behavior geneticists' historical narratives, members of the lay public were either too genetically deterministic or too focused on free will, but rarely were they just right. But unlike those molecular biologists, nonscientists also held differing views on the justification for animal research and the biomedical model of mental illness. Trying to balance these multiple, competing concerns left researchers at Coast feeling very insecure about how to navigate public communication. Within the boundaries of the scientific community, researchers at Coast could err on the side of caution and make conservative statements that downplayed the significance of their genetic findings. But when talking to nonscientists, they believed that too much caution could backfire, leaving the claims of animal rights activists and deterministic or stigmatizing assumptions about mental illness unchallenged.

COMMUNICATIVE CULTURE CLASHES

It was in precisely the settings that practitioners felt the greatest need to manage their claims that they simultaneously experienced the breakdown of their most effective tools for doing so. Researchers at Coast already found it difficult to calibrate the scope and certainty of their claims for the scientific community, but these difficulties were intensified by the conflicting imperatives of public communication. Under heightened pressure to create claims of just the right formulation, they reached for the familiar techniques that they used to signal their epistemic commitments to each other, and found them to be largely inadequate for talking to nonscientists. As I will show, many of these techniques were in direct conflict with the conventions of science reporting, reinforcing researchers' impressions of a communicative gap between themselves and journalists.

Speculating on Future Applications

I argued in chapter 2 that one of the ways that Coast researchers tried to moderate their claims was by extending the time line of their research. Rather than making five- to ten-year forecasts, they placed likely applications of their research to clinical practice in the far distant future. By doing so, they avoided overpromising in a world saturated by genohype and guarded against the conclusion that there were simple solutions available for what they believed were complex problems. Journalistic conventions, in contrast, tend to emphasize the clinical importance of scientific discoveries and compress time lines to application. Jane Gregory and Steve Miller (1998) identify "meaningfulness" and "relevance" as two aspects of what makes a science story valuable and newsworthy to a journalist. Emphasizing the potential applications of biomedical findings helps journalists satisfy both of these values at once: discussing clinical application helps readers understand the importance of the finding and see it as relevant to their own lives.

These opposing orientations were evident in how Coast researchers perceived the sequence of questions that journalists typically posed to them. Against the backdrop of the extended future they envisioned for their field, researchers thought that journalists made overly fast leaps from background information directly to questions about treatment. Dr. Martin described her experience to me as follows: "They want to know the basics, you know, how is methamphetamine made? Why is it so addictive? If you do identify genes that are involved, what will that do for us? How do you envision this helping? Are you—they always of course want to know if you're heading towards, is it a treatment so that you can come up with drugs, or is it genetic engineering? Do you plan on changing people's genes in order to help?" In Dr. Martin's view, in a typical media interview about her research, she spent almost no time talking about the research itself. She was happy to provide information on the larger context of methamphetamine addiction and give her thoughts on what treatment might look like in the future, but she was worried that readers might walk away with the idea that new treatments were around the corner.

To help them answer these types of questions, many researchers at Coast took a media training course with Eric, a local television and radio personality who offered one-on-one coaching for scientists. Eric had a makeshift studio on campus, where he had installed a green screen, lights, and video camera in one of the hospital conference rooms to simulate a television interview experience. When I visited, Eric was working with Dr. George Taylor, a senior scientist with wiry white hair who squinted uncomfortably under the bright lights. As he answered Eric's questions, he looked off toward the upper corner of the room, seemingly trying to ignore the large camera lens positioned inches from his face. Eric was coaching him in the art of "relaxed intensity"—Eric's term for speaking with "focus and clarity" while maintaining a "conversational" tone. Eric had instructed Dr. Taylor to think of three messages that he wanted to communicate in his interview and to try to convey those messages in his answers no matter what questions he was asked. After asking a few questions and listening to the responses, Eric tried to guess Dr. Taylor's messages. Eric's first guess was the following: children exposed to methamphetamine during pregnancy are more likely to have memory problems. "Almost," Dr. Taylor replied—his message was that a particular neurotransmitter mediated the cognitive effects of methamphetamine exposure. The difference between the two messages struck me as so pronounced that I had to stifle a surprised laugh.

Discussing the training session with me afterward, Eric told me that this was a common occurrence. The problem, in his view, was not that scientists were unable craft a straightforward message about their research but that the kinds of messages they wanted to deliver were not the kinds of messages journalists wanted from them. Most researchers could offer an explanation of the clinical relevance of their research, he said, but relatively few offered such explanations without being asked directly. He told me about how he had been working with Dr. Taylor on some messages about the implications of his research that Dr. Taylor could deliver in public. But as Dr. Taylor chimed in on the training session postmortem, it became clear to me that their goals were not as aligned as they first seemed. Dr. Taylor's laboratory used both human brain imaging studies and animal models of learning in their research on drug addiction and memory, and he told me that when asked to speak about his work in public, he typically chose to talk about his mouse research. When he talked about his imaging research with children, his experience was that audience members immediately started relating his work to their personal lives, often in ways that he felt were overly fatalistic. He felt keenly aware that when he talked about his research that there might be people in the audience with relatives struggling with alcoholism or who had adopted children born to drug users (a colleague of his, he said, had recently adopted a baby with prenatal methamphetamine exposure). But when he talked about his mouse models, he found that audience members asked fewer questions about their relatives or children. Presenting on animal research rather than clinical studies seemed to me a technique for resisting the public communication imperative to talk about application, because Dr. Taylor found that nonscientists viewed the mouse models as less relevant to their lives.

Others resisted the push to talk about application more directly. In the session on ethics that I moderated for the introductory behavior genetics class, much of the discussion focused on worrisome future scenarios, such as prenatal genetic screening for common diseases and the possibility of selective abortion. Dr. Lam, Dr. Smith's postdoc, was impatient with this line of conversation. The complexity of these diseases and their underlying genetics meant it was unlikely that any genetic tests of this kind would be forthcoming, he argued. "Animal behavior genetics research has gained a lot of detail in describing what's abnormal," he said, "but that information doesn't have a lot of predictive value." To talk about prenatal screening, then, was to misunderstand the state of the science. As the conversation went on, he became more visibly frustrated. "Look," he finally said, "for a kid born today, we have no more ability to predict his future by looking into his genome than a Mayan priest did by looking into a fire!" In comparing the current state of animal behavior geneticists' knowledge to that of supposedly primitive societies, Dr. Lam resisted the orientation of ethics discussions toward future applications and attempted to re-extend the field's time line so that speculations about application seemed unwarranted.

Eliminating Cautious Language

Another technique that researchers at Coast relied on to modulate their claims was using specific phrases that signaled their conservative epistemic commitments. As I argued in chapter 3, the distinction between describing the elevated plus maze as a "test of anxiety" versus a "test of anxiety-like behavior" was important to researchers, because for them, these phrases reflected different understandings of the knowledge production capacities of the test. But the complexity talk and cautious language that was ubiquitous at Coast suddenly became problematic when they spoke to nonscientists. Phrases such as "anxiety-like behaviors" or "tests of antidepressant efficacy" simply sounded like technical jargon and failed to have the intended effect of preserving some distance between researchers' mouse experiments and the human behaviors they were intended to model.

If "complex" was one of the most frequently used adjectives in the laboratories at Coast, then "simple" was its counterpart in media training sessions. "One of the most difficult things about dealing with scientists is that they don't even know they're talking shop," Eric told me. "They actually think that they're being simple." During the training session, I watched as Eric repeatedly called out Dr. Taylor on his use of the word "neurotransmitter" as too jargony. Eric suggested that Dr. Taylor substitute the phrase "brain chemical" for "neurotransmitter." Dr. Taylor thought this sounded strange, but he persevered. He seemed less comfortable, however, with other substitutions. Eric wanted to hear about memory problems in general, while Dr. Taylor wanted to talk about specific types of memory—spatial memory, long-term memory, or fear conditioning. Substituting "memory" for "spatial memory" was not the same to Dr. Taylor as using a more colloquial phrase to talk about neurotransmitters. For Dr. Taylor, each of these was a distinct type of memory, involving different brain structures and molecular mechanisms. Eric's suggestion to speak about memory in general had the unintended effect of inflating the scope of Dr. Taylor's claims, making it sound as though his results applied to many different types of memory rather than just one.

Other well-meaning exhortations from journalists and media specialists to use "simpler" phrases similarly conflicted with researchers' local conventions for talking about their work. Dr. Smith told me that he sometimes called out journalists when they asked him what his research had "proved," telling them that he had not proved anything, only failed to disprove the null hypothesis. I could see how for Dr. Smith, this distinction was meaningful. Hewing to the language of the scientific method—where a theory can always be disproven and no amount of data can ever prove it to be true—was a way for him to emphasize the tentative and unfinished nature of scientific work. But I could also see how to the reporter, this distinction likely seemed pedantic. Such overly formal descriptions would not help nonscientists to grasp the significance of the research.

Stripped of their linguistic conventions for managing the scope and certainty of their claims, practitioners at Coast found it difficult to craft statements that they felt comfortable with. Graduate student Hannah described her struggles with public communication as follows:

> We really don't know very much, and so I find it very hard to come up with a definitive answer for myself or to say–even if it's a simplistic answer. You know, something like, well, we think alcohol works through this system. I feel like I need to be like it could also be this, this, this, and this, and this this this—so what I really need to improve on is the—almost the PR statement. Like, this tells us this. That's not the whole truth, it's not everything, but here's a good statement to take away from this.

Clarity and unambiguity is another journalistic value that Gregory and Miller (1998) argue informs science news reporting, and it is another value that is in direct conflict with cultural conventions at Coast. Hannah's socialization had taught her that unqualified statements were to be avoided at all costs. Even as she attempted to formulate a hypothetical "PR statement" for me about her research, she also modulated it by inserting the phrase "we think" before her statement about alcoholism, and following that tentative hypothesis with the qualification that it was "not the whole truth." The more that journalists and media trainers tried to eradicate phrases like these from Coast researchers' speech, the more researchers felt as though they literally did not have the words to express themselves to nonscientists.

Extrapolating from Controlled Settings

Adjusting how they talked about their research was not the only way that practitioners at Coast attempted to control their claims; they also adjusted their experimental practices with the complexity of behavior in mind. As I explored in chapter 1, controlling the experimental setting was a technique that they relied on heavily to manage the multitude of environmental factors they believed could alter behavior. Because they assumed that the impact of a single genetic alteration was likely quite small, they assumed that they needed a highly uniform environment to make those small effects detectable.

Researchers at Coast saw the ability to create exquisitely controlled environments as one of the advantages of working with mouse models, but when it came to public communication, this strength transformed into a weakness. Highly controlled animal experiments lent themselves to "gene for" stories because of how they isolated individual factors. In the news stories reporting on animal behavior genetics research that I examined, the much reviled "gene for" trope rarely appeared in either researchers' quotes or journalists' descriptions, but this message was often implicit in the narratives. Take, for example, the opening paragraphs of a short article reporting on a knockout study on anxiety: "As if life wasn't nerve-racking enough, scientists have created a strain of mice that are fraught with anxiety. Audrey F. Seasholtz of the University of Michigan Medical School created the mice by deleting a gene that controls production of a protein called corticotropin-releasing hormone (CRH), which is involved in regulating stress. Male mice without the CRH gene seemed much more anxious—they were less likely to leave a protected chamber or explore open areas, the researchers report in the Sept. 28 issue of the Proceedings of the National Academy of Sciences" ("Science Notebook: Mice with High Anxiety" 1999). Although the article makes no claim that the CRH gene is a "gene for" anxiety, all of the explanatory power for the sudden change in the mouse's behavior rests with the gene in this story. This is not a misrepresentation of the experiment per se—knockout experiments are designed to isolate and study the influence of individual factors on psychiatric disorders. Rather, it is what Sharon Dunwoody (1982) has called a subjective inaccuracy, one where the error that scientists identify lies not in inaccuracies in fact but inaccuracies in meaning. Researchers at Coast objected to stories like these because first, they extracted individual results from ongoing research programs designed to investigate multiple genetic factors, and second, reporters gave no weight to environmental factors. Without this contextual information, they believed nonscientists were likely to take away a reductionist message about anxiety research. Researchers saw similar problems with the public presentation of other kinds of animal studies, such as those examining the role of individual neurotransmitters or brain regions in behavior. Here as well, they worried that the lay public was likely to miss the message that there were multiple brain systems involved in most behavioral disorders.

Comparing the CRH story quoted above with stories reporting on human behavior genetics studies makes it clear how the controlled form of animal research worked against researchers in science news settings. Human researchers have far less control over the biology of their research subjects than animal researchers, and consequently they rely heavily on statistical techniques to parse the different factors at work in the development of psychiatric disorders. Coast researchers viewed this as one of the disadvantages of doing human research, but in public communication, this transformed into an advantage. These techniques, by their very nature, tend to emphasize the multifactorial nature of behavioral disorders. For example, genome-wide association studies (GWAS), one of the favored tools of human behavior genetics, use statistical techniques to scan the genomes of large numbers of affected and unaffected people. The data from GWAS are typically displayed in what are known as

"Manhattan" plots, named for their resemblance to the Manhattan skyline. In these images, data on locations spanning the entire genome are displayed together, with the statistically significant regions of the genome towering like skyscrapers over those that are not linked to the disorder. Rather than presenting a single gene in isolation, these images depict genomic findings in context, showing multiple significant genomic locations at once alongside other locations with varying degrees of association with the disorder under investigation.

When journalists write about the results of these studies in news media articles, complexity and the multifactorial nature of psychiatric diseases are prevalent themes. The simple fact that GWA studies typically report on multiple genes rather than discussing a single gene at a time alters narratives about the etiology of psychiatric disorders. A news brief on a study on schizophrenia described the outcome as follows: "The study revealed five new genetic signals that were important in the development of schizophrenia, or the risk of developing it. It also confirmed the previous discovery of another two genetic signals. 'It clearly demonstrates that schizophrenia has behind it a complex genetic inheritance pattern,' Professor Scott said. 'It's not a transmission from one generation to the next. It requires a number of genes to be inherited. We don't know what that number is. This study shows there's at least seven involved'" (Jones 2011). This description of schizophrenia as a disorder with at least seven genes contributing to its development presents a much different picture of gene action than similarly brief reports on knockout studies. The statement that schizophrenia requires a number of genes to be inherited leaves open the question of what might happen if an individual inherited only two or three of the disease-associated genomic variants, implying a model of genomic risk rather than deterministic causation.

Treating multiple genes together also created opportunities for researchers to push back against the imperative to talk about future treatment. In contrast to the straightforward speculations about clinical applications often found in news media articles reporting on animal studies, stories about GWAS often painted a more complicated picture of the clinical future. An article reporting on another schizophrenia study, for example, contained optimistic speculations about new treatments but also included some warnings: "Should most of the genetic component of the disease turn out to depend on multiple rare variants," one journalist wrote, "the task of finding general treatments might seem to be far harder than if a few common variants were involved" (Wade 2008).

Filtering Out Epistemic By-Products

Finally, as I argued in chapter 4, researchers developed and sustained their understandings of complexity by preserving the epistemic by-products they gained through experimenting in the laboratory. Working with genetically identical animals that differed in their behavior reminded scientists about the limitations of genetic explanations of psychiatric disorders, and researchers at Coast preserved and circulated such observations of difference to intervene in their fellow researchers' understandings of complexity and gene action. Science news articles reporting on their research, however, devoted little space to these kinds of observations. News articles contain an implicit hierarchy of reportable information, one that prioritizes the main finding of a published paper and its implications for human health and only includes discussions of complicating factors, caveats, or anecdotes as space allows. The omission of this information once again contributed to Coast researchers' impression that journalists and the news articles they produce were determinist or reductionist. Because the experience of accumulating epistemic by-products was an important means through which researchers at Coast developed their sense of complexity, they believed that eliminating this knowledge would lead the public toward an overly simplistic view of behavior.

One way of making this implicit hierarchy visible is to compare different lengths of newspaper articles reporting on the same finding. A series of articles reporting on a 1998 *Nature* paper provides an instructive example of this hierarchy (Thiele et al. 1998). The study on the effects of knocking out the gene coding for neuropeptide Y received international news media attention, and the short news articles in particular show a remarkable degree of consistency in the type of information they present. One article covering the publication, from the Australian daily *The Courier Mail*, is short enough to reproduce in its entirety:

> Scientists may have found a biological reason why some people crave alcohol more than others. Researchers at the University of Washington have found a link between the level of a particular molecule in the brain of a mouse and its drinking habits. By genetically altering the levels of a molecule in the brains of mice, the researchers discovered high levels of the molecule, known as neuropeptide Y, made the mice drink more alcohol. Two groups of genetically altered mice were allowed to choose between drinking water and a wine-strength cocktail of ethanol and water. The re-

> searchers found mice with no neuropeptide Y were more partial to alcohol than water. Mice with extra neuropeptide Y tended to consume less alcohol and actually fell asleep afterwards, which the researchers determined was due to their sensitivity to alcohol. The researchers believe their discovery could benefit companies looking to develop drugs to combat alcoholism in humans. ("Brain Link to Taste for Drink" 1998)

An article published in the *Washington Post* contained largely the same information, although it used quotes from the researchers to summarize the findings and elaborated on their potential implications for human health. But from this slightly longer article, we also learn that neuropeptide Y was identified a decade and a half earlier in studies of obesity and depression ("Findings: Alcoholism and Neuropeptide Y" 1998). What the article suggests but does not explicitly state is that the gene in question plays a role in multiple behaviors and physical characteristics, including alcohol preference. It is not until we get to a 550-word article reporting on the study, however, that a discussion of complexity, the environment, and the limitations of the findings appears. This article, appearing in the *Philadelphia Inquirer*, begins with a discussion of the findings and their implications for science and human health that resembles the *Washington Post*'s report. But three hundred words in, the article begins to include some of the epistemic by-products researchers generated through this study:

> Animals lacking neuropeptide Y were more anxious than normal mice in the study. But animals with extra neuropeptide Y weren't calmer, Thiele said. In addition, all mice in the study ate the same amount and weighed the same, suggesting that any links among stress, eating and drinking were likely to be complex. The finding injects hope into the seemingly daunting task of tracing the genetic roots of alcoholism in people. The complex disease is shaped by psychological and environmental factors and also, according to evidence from studies of alcoholism among families and twins, by genes. But the much-publicized 1990 discovery of a human "alcoholism gene"—one that influenced how the brain responds to a neurotransmitter called dopamine—has been largely discredited. Scientists now believe it is unlikely that a single gene increases susceptibility to alcoholism. They think a suite of genes works together to influence various aspects of alcohol use, such as craving, tolerance and withdrawal. It is also possible that different subtypes of alcoholics have differing underlying genetic mechanisms. (Lee McFarling 1998)

This version of the story noted the surprising lack of variation in the body weights of the mice—something that was not the focus of this particular study but that researchers expected they should see based on previous studies. The news article pivots from this epistemic by-product to a discussion of other surprising reversals of previously established findings and concludes by discussing scientists' understandings of alcoholism as a multigenic disorder.

Dr. Smith particularly liked news articles like these that, in his words, communicated the "process" of science rather than simply the outcomes of experiments. He had adopted this language from a *Science* magazine letter, titled "Going Public with the Scientific Process," which was published during my stay at Coast (Cronje 2008).[5] The letter argued that "confining science messages to just the facts interferes with public understanding of science as a systematic, logical process of human inquiry" (1483). Instead of reporting only on currently accepted findings and "sweeping uncertainty under the rug" (1483), the author argued that science communicators should aim to represent the dynamic process of science, whereby facts were refined or altered over time. Dr. Smith thought that this was a good diagnosis of the problem with science reporting and clipped the article from his paper copy of the journal to show me at one of our weekly meetings.

As I argued in chapter 4, however, the distinction between the process of science and the facts it produces is not always clear. Researchers could and sometimes did transform by-products of their inquiries into products of their own, and when they did, the news media stories reporting on those articles conveyed quite different messages about behavior. For example, the multi-sited study discussed in chapter 4 (which aimed to quantify the effect of the laboratory environment on mouse behavior) generated a number of news articles, many of which portrayed behavior as environmentally rather than genetically determined. One article from the Canadian newspaper *The Globe and Mail* reported on the study finding that mice in the Edmonton laboratory behaved differently than mice in the other two laboratories when given cocaine and when tested on the elevated plus maze. The article opened by asking, "Is it the water or the clean prairie air that makes mice in Edmonton more relaxed and less interested in drugs than their U.S. cousins?" (Immen 1999). By (presumably jokingly) explaining this difference in terms of the water the mice were drinking or the "clean prairie air," the article gave the impression that anxiety and drug use are shaped by environmental factors. This message is in conflict with popular conceptions of these disorders as either genetically determined or as "diseases of choice," and therefore seems to be in conflict with the journalistic norm of "consonance" that Gregory and Miller (1998) describe. But rather than seeing this study as newsworthy because it aligned with existing beliefs and attitudes, journalists seemed to find it newsworthy precisely because it did not. Another news article similarly played off this reversal of expectations, succinctly summarizing the lesson of the study with the headline "Mice Study Shows Genes Are Not Always Destiny" (1999).

By transforming epistemic by-products into publishable findings that journalists found newsworthy, researchers could elaborate on their understandings of gene action and voice their concerns about the limitations of existing experimental techniques. One article reporting on follow-up research to the multisited study interviewed Douglas Wahlsten, one of the original study authors, and quoted him at length about his views:

> "We cannot understand what genes do," Wahlsten says, "without knowing how the environment works, and the environment is even more complex than the genome." . . . Rather than trying to eliminate them, the researchers are trying to find ways to precisely document and quantify these environmental effects. Once they can do that, then they can ask what makes a particular strain sensitive to them. That, he says, would help researchers get closer to the gene–environment interaction, which many in the field believe shapes behaviour. "We've known for a long time that there's no gene anywhere that codes for a specific behaviour," Wahlsten says. (Ogle 2002)

Wahlsten's intricate message about gene–environment interaction is not one that regularly appears in news reports on animal behavior geneticists research, but this article suggests that there is no in principle reason why such information cannot be considered newsworthy and communicable to the lay public.

Studies that attempted to quantify and circulate environmental knowledge were rare, however, and perhaps consequentially, so were discussions of this knowledge in news media reports. This should not be understood as merely a problem of journalistic attention: the reasons why journalists attend to some kinds of information and not others are manifold, and journalists alone are hardly responsible for creating the hierarchies that allow some information to travel more easily than others. It is the animal behavior geneticists who first create distinctions between products and by-products that result in asymmetries in the published literature, which are then reproduced and amplified by journalists. Journalists were often simply replicating the hierarchies between genetic and nongenetic findings that practitioners themselves had established. But the further loss of epistemic by-products that took place as journalists distilled animal behavior genetics research down to popular narratives contributed to researchers' impression that journalism distorted the meaning of their work.

REPAIRING UNDERSTANDINGS OF GENE ACTION

To summarize, many aspects of the research culture at Coast that I have described throughout this book seem at odds with the conventions and values of journalism. The extended futures that Coast researchers created, the cautious language that they used, and the epistemic by-products that they preserved were all incompatible with how journalists expected them to speak about their research. The results of their highly controlled laboratory work, which they valued for its potential capacity to tease out associations from complex systems, appeared to them to be determinist and reductionist when displayed in a format designed for individual, finished findings. Journalists' hierarchies amplified their field's own implicit hierarchies of what counted as interesting and publishable findings, further concentrating the focus on the role of genes in behavior.

These numerous cultural clashes contributed to researchers' subjective perceptions that the news media was a place where they had little control. Dr. Smith, for example, was very clear on the message about the complexity of behavior that he wanted to convey in media interviews, but he felt that more often than not, that message did not get through. He told me, "The lay public needs to be educated that genes do not determine behavior, and that environments don't determine behavior, but that genes predispose to more or less of a behavior and environment also does the same thing, and the interaction is what determines the behaviors. So yeah, every interview I've ever given a reporter has probably asked that question, and I've given them that answer. We just have to keep doing it until people understand it." Researchers experienced public communication as a site of frequent communicative breakdowns between themselves and their interlocutors, and the strategies that they deployed in media interviews or conversations with family members reflected this understanding.

Dr. Smith's repetition of the gene–environment interaction message can be understood as a form of what conversation analysts call "repair" work—a set of techniques that interlocutors use when one senses that there are problems in speaking, hearing, or understanding in a conversation (Sacks, Schegloff, and Jefferson 1974; Schegloff, Jefferson, and Sacks 1977). Conversation analysts have argued that participants in a conversation frequently locate potential "trouble sources" for each other—such as confusion about whose turn it is to speak next—so that the communicative problems can be fixed. When successful, repair work unobtrusively restores order to the conversation so that it can proceed smoothly. For researchers at Coast, the main "trouble source" in their conversations with journalists or family members was communicating their understandings of gene action. Researchers might describe a gene as "influencing" behavior, intending to convey that the gene under discussion was only one factor of many and that its mechanism of action was not fully known. Their conversational partners, however, might reply in a way that indicated that they took away the message that the gene was the determining factor.

News media articles show evidence of researchers' attempts to repair these misunderstandings and draw their interlocutors into a shared understanding of gene action. For example, take a *New York Times* article on animal models of alcoholism from 1995—still early enough in the days of the "gene finding" revolution that the article could proclaim that "more than two dozen groups of researchers are plunging into this new field" (Kolata 1995). The journalist interviewed Lee Silver, a Princeton University geneticist whose description of his research seems at first to ascribe much causal power to genes. "Let's take alcohol," Silver is quoted as saying, "I can look at any strain and tell you exactly how much this animal will drink—some drink in moderation, some drink in excess, and some don't drink at all." But Silver also addressed bluntly misconceptions about gene action. The article quotes him as saying, "I disagree with people who say that genes are everything in humans. . . . Those people don't understand a couple of notions. Even if you're predisposed, that just means that your probability of having the behavior is greater. All these genes do is increase predispositions. And human behavior, if anything, has to be under an enormous number of influences. I'm as angry with those who say it's all genetic as I am with those who say it's not genetic at all" (Kolata 1995).

The article included a series of other cautionary quotes from historians, medical anthropologists, and fellow scientists. But it is worth noting that the article quotes Silver himself in ways that both reinforce and challenge the role of genes in behavior. One can imagine how these seemingly contradictory quotes might have emerged in the course of an ongoing interaction: Silver's explicitly antideterminist statement quoted above could be an attempt to correct what he saw as a misunderstanding about the role of inherited factors in alcoholism and clarify his position for the journalist.

Other indications of researchers' attempts to repair reductionist or determinist narratives are sprinkled throughout news articles. Here is a sampling of the different kinds of counterbalancing statements that appear in news articles covering rodent studies of behavior and psychiatric disorders:

> "Cannulas in the brain may be fine for animals, but not for 4 million Alzheimer's patients," said Baldino of Cephalon. "It's a great way to show proof of concept—to show that growth factors can have an effect in the brain. But practically speaking, I don't think patients are going to be lining up at the clinic." ("Diseases That Attack the Brain" 1995)

> As enthusiastic as Dr. Zhang is about the potential health benefits, he warns against running out for a toke in a bid to beef up brain power or calm nerves. "There's a big gap between rats and humans," Dr. Zhang points out. (Walton 2005)

> Dr. Joe Z. Tsien, a professor of molecular biology at Princeton who genetically engineered smarter mice a few years ago, says he is skeptical that the results can be transferred to people. "If you look at how people improve their brain power, it's through education," he said. "That has proven to have 100 percent efficiency with minimal side effects." (Pollack 2003)

> Flint cautioned that nurture still dominates. The genetic component probably accounts for just 20 percent of variability seen in the mice. "The environmental determinant is by far the greater," Flint said. (Schwartz 1995)

> But don't count on the discovery changing your life any time soon. Any treatment is said to be 5 to 10 years away. And the field of genetic disease is littered with false hopes gone unrealized. It is humbling to recall the excitement that greeted past announcements of genetic defects associated with alcoholism, cancer, manic-depression, schizophrenia and other ailments. Only later did scientists realize they had either misidentified the gene or else had no idea how it operated, and thus no idea how to devise a treatment. ("The Obesity Gene" 1994)

Each of these statements manipulates a different parameter of scientific claims making, but they all serve to diminish the potency of genetic findings. Some statements alter the scope of the claims themselves, reducing them from a statement about mammals to a statement about mice. Some statements extend the time line of the research program and insert distance between the existing findings and clinical application. Others draw attention to existing knowledge about environmental factors, counterbalancing messages about the importance of genes and instead insisting that behaviors are complex phenomena with multiple inputs. Other passages explicitly highlight past failures in the field as a way of reducing the perceived long-term stability of present genetic findings. That these caveats appear repeatedly in published news media articles is evidence both of scientists' power to control the framing of reports on their research and also of their subjective sense of powerlessness. Repeatedly correcting what they perceived as breakdowns in communication with their interlocutors did not leave researchers at Coast with the impression that they were in control of media frames about their research.

Engaging in attempts to repair perceived miscommunications about the nature of gene action is not without risk for scientists. Erving Goffman's (1955) research on "face-work" emphasized how social status, and not just social order, is at stake in such conversational exchanges. Goffman described face-work as the verbal or nonverbal actions that individuals take in a social encounter to maintain a consistent, positive self-image. The concept of face-work shares several commonalities with conversation analysts' description of how interlocutors maintain order in their interactions. Interlocutors generally work together to sustain each other's self-image, Goffman argued, and use repair work (or in his terms, corrective processes) to indicate lapses and provide opportunities to correct them. He argued that such work is necessary not just to maintain order in social interactions but also to keep one or more participants from suffering a "loss of face"—an embarrassing diminishment of respect and social standing.

In some situations, repair work may accomplish both goals of allowing communicative exchanges to proceed smoothly and permitting interlocutors to maintain a positive self-image. But animal behavior geneticists' media interactions suggest that these two aims are not always aligned. In this case, practitioners' strategies for repairing what they saw as misunderstandings about gene action seemed to come at the expense of the field's "face," trading mutual understanding against a positive image for the field. Not all expressions of uncertainty or antideterminist sentiments may diminish the field's social standing (and as I argued in chapter 2, expressing some degree of epistemic modesty may actually enhance it), but making too many caveats risked rendering the practitioner or the field's face incompatible with the accepted self-image of science. As Dr. Smith put it to me, he often felt when he was trying to carefully formulate his claims in public venues that he looked "wishy-washy" and "unscientific." Researchers feared that if they reduced the explanatory power of genes or mouse experiments too much, they might open themselves up to questions about whether their research was actually valuable, or whether what they were doing was really science.

News media coverage of the multisited study on the laboratory environment illustrates the difficulty of accomplishing both of these goals at once. As I argued above, this study provided opportunities for researchers to speak publicly about the process of science and the importance of environmental factors in behavior, but it also opened researchers up to criticism of their field. *Science* magazine published a news brief alongside the original study titled "Fickle Mice Highlight Test Problems" (Enserink 1999). "Studying the genetics of behavior is like riding a roller coaster," the article began. "No sooner has one group of researchers tied a gene to a behavior when along comes the next study, proving that the link is spurious or even that the gene in question has exactly the opposite effect" (1599). The *Globe and Mail* article on the study similarly called into question the stability of behavior genetics research. The article ran with the headline, "Canada's Relaxed Mice Puzzling," followed by a subtitle, "Differing Behavior Could Nullify Experiments," which directly challenged validity of the field's findings (Immen 1999).

The study authors offered their own analysis of media coverage of their study in an article following up on their original results and took issue with the slant of both of these news pieces. They disliked the *Science* magazine article's "pejorative" headline about fickle mice, and they described the *Globe and Mail* headline as "appalling" (Wahlsten et al. 2003, 306). The study authors also took issue with an essay discussing their research that appeared in the *New Yorker* several years later, which used the multisited study as an example of how established scientific facts seemed to be "losing their truth" (Lehrer 2010).[6] One of the authors wrote a letter to the editor of the *New Yorker* in response to this piece, voicing his concern that nonscientists might take away the message from the article that "science is a useless exercise" (Hughes 2012).[7]

Engaging in repair work, then, did not always result in a positive self-image for the field, even when that repair work was successful. When animal behavior geneticists restricted the scope of their claims, drew attention to uncertainties in their methods, or introduced environmental knowledge to counterbalance their genetic findings, they risked upsetting popular understandings of science as cumulative and certain. When news reports described the trajectory of behavior genetics research as a "roller coaster" or researchers' favored model organisms as "fickle," practitioners worried that these articles would lead nonscientists to question whether behavior genetics was a mature scientific field, or whether it was scientific at all.

CONCLUSION

That scientists view public communication as difficult and dangerous is a commonplace statement, so much so that it may seem unnecessary to even ask why this is the case in animal behavior genetics. STS and science communication scholars have tended to argue that scientists enjoy a great deal of authority in shaping news media coverage of their research, and scientists' complaints about the press could simply be seen as one of the means through which they exercise this authority. Publicly decrying news articles as genetically determinist or reductionist is a way for animal behavior geneticists to establish what counts as a balanced representation of their research, and perhaps also to enhance their reputations as careful scientists. But I have suggested in this chapter that it also worth taking practitioners' complaints seriously on their own terms, and asking why it is that they experience a lack of agency in public communication even when they hold so much of it.

By exploring in depth Coast researchers' culture of claims making, we can see more clearly why they felt powerless when speaking in public venues. While they may have had university press offices or embargos imposed by journals to control the flow of information on their behalf, the techniques that they relied on most heavily to manage claims within their scientific community seemed to be ineffective outside of it. In particular, it was the techniques they used to restrict claims—using linguistic markers of cautious epistemic positions or counterbalancing genetic findings with environmental knowledge gleaned through work in the laboratory—that were especially ineffective in popular venues, not the ones they used to make claims seem more authoritative and certain.

One potential reason these disconnects between journalistic cultures and scientific ones have not been especially evident in STS work is that scholars have focused more on the erasure of uncertainty than they have on its preservation. Latour's (1987) and Susan Leigh Star's (1985) classic accounts of laboratory work, for example, describe the mechanisms through which scientists transform "local uncertainties into global certainty," as Star (1985, 391) has put it. Sara Delamont and Paul Atkinson's (2001) more recent study of graduate students similarly describes how aspiring scientists are trained to eliminate uncertainties, tacit knowledge, and local contingencies from their public presentations. Monika Cwiartka's (2011) work on mouse experiments likewise examines specifically the rhetorical techniques involved in transforming the local, variable bodies of mice into persuasive arguments about behavior.

Studies that have examined the preservation of uncertainty in public venues have tended to emphasize similarities between how scientists manage uncertainty in scientific and popular settings. Brian Campbell's (1985) and Stephen Zehr's (1999) descriptions of how experts use uncertainty to discredit the claims of competing groups, for example, resemble Latour's (1987) adversarial account of how battles are waged within scientific communities. These accounts, while valuable for drawing attention to the maintenance of uncertainty in public venues, do not capture differences between the processes taking place inside and outside of the laboratory—in these cases the preservation of uncertainty serves similar functions and is accomplished through similar processes in both scientific and popular venues. Christian Greiffenhagen and Wes Sharrock's (2011) study of mathematics takes a unique approach to studying uncertainty in different venues, attending to how practitioners re-create the uncertainty that has been written out of textbooks or scientific papers as they use these objects in practice. They frame their study as a challenge to the idea that proofs and papers present an overly certain view of mathematics, an idea that they note has been promulgated by mathematicians themselves. But rather than exploring why practitioners express concern about how their papers might be read by outside audiences, Greiffenhagen and Sharrock are focused on discrediting this "myth" (860) by deconstructing the distinction between the certain "front" and fallible "back" stages of mathematics.

The processes by which global certainties are created are certainly deserving of analytical attention, especially because erasing the active processes of scientific construction is one means by which science gains and maintains its cultural authority. So, too, are the processes by which the scope of scientific claims are inflated and by which environmental factors are erased and genetic findings left to stand unchallenged. But we should neither assume that scientists and analysts are always pushing on the opposite sides of these levers, nor should we lose sight of the fact that there is no independent position from which to evaluate the appropriate scope, degree of certainty, or degree of determinism in a scientific statement. Taking seriously the concerns that scientists express about public communication might even open up opportunities for interventionist projects that are based on shared commitments rather than oppositional positions. Analysts and actors might find themselves aligned through their mutual dissatisfaction with popular accounts of behavior genetics research in a shared program of attempting to "emancipate the public from prematurely naturalized objectified facts," as Latour (2004, 227) has so eloquently put it. By sitting with the paradox that is researchers' discomfort with narratives that are in many ways of their own making, we may collectively find more satisfying ways of remaking them.

CONCLUSION

An Expanded Vocabulary for the Laboratory

My aim in this book has been to explain in depth how one particular group of researchers carries out the work of animal modeling. This study offers a useful contrast to existing science and technology studies (STS) work because of how the researchers I studied combined an orientation toward complexity with the work practices of the laboratory. Rather than studying human populations in all their messiness or adopting the tools of big data to manage multiple, interacting factors, they chose to remain within the highly controlled space of the laboratory. Their decision to work with animal models and study specific genes, brain regions, and facets of pathological behaviors was not a rejection of complexity; it was a different way of engaging with it. Through their experimental work, they aimed to produce findings that resembled the partial, indeterminate, pragmatic claims of scientists working in field sites or clinical settings. By merging these two sensibilities, this particular epistemic community expands our understanding of what laboratory research is capable of producing.

In focusing intently on the way that this community works, I have said relatively little about how they might have come to operate in this way. In the conclusion, I will broaden my focus and situate the Coast community in some larger disciplinary tensions and historical trends. From this more expansive vantage point, I aim to offer insight into both the unique combination of factors that shaped the Coast laboratories and how we might apply the arguments I have developed through examining this particular site to understand other locations.

There are three sets of factors that are important for understanding the specificities of experimental life at Coast. The first is its position in various scientific fields. Researchers at Coast identified with multiple disciplinary communities: for example, Dr. Smith, who trained as a psychologist, maintained a membership in the Behavior Genetics Association (BGA) and regularly attended neuroscience and addiction research meetings. Considering Coast researchers' position in these various fields and the position of these fields in relation to each other helps explain their intense concern with methodology.

In the alcohol and addiction research fields, researchers at Coast were operating from a position of strength. Both the past and present directors of the National Institute on Alcohol Abuse and Alcoholism (Ting-Kai Li and George Koob, respectively) are researchers who have worked with rodent models and have even been involved in Alcohol Research Group (ARG) initiatives. While animal researchers are still called on to justify what some see as an overinvestment in biomedical approaches to understanding addiction (Hall, Carter, and Forlini 2015; Volkow and Koob 2015), broadly speaking their preferred methods have received strong support in addiction research over the past several decades. The same cannot be said, however, of Coast researchers' position in other fields. As Panofsky's (2014) research has shown, animal researchers occupy a relatively marginal position in contemporary behavior genetics. While animal research was central to the field as it was established at midcentury, human studies came to dominate venues such as the BGA meetings in the latter half of the twentieth century. The epistemological and ethical sensibilities expressed at Coast are also underrepresented in the venues most closely associated with behavior genetics today. For instance, Panofsky (2014) argues that the majority of behavior geneticists responded to the Jensen controversy by defending his right to make provocative scientific claims and discouraging those within the field from making methodological critiques that resembled those made by Jensen's attackers—moves that, as we have seen, are the opposite of the mantra of cautious claiming adopted by researchers at Coast.

Coast practitioners fared only a little better in the broader neuroscience field or with respect to researchers from other disciplinary backgrounds who also claimed to study the genetics of behavior. Coast researchers felt that they were at a disadvantage in the "molecular revolution" of the 1990s compared with the molecular biologists who had newly turned toward behavior. They described this as a problem of the status and of the material culture of their respective disciplines: behaviorists found it difficult to exert control over their techniques because the barriers to participate in behavioral research were low. For example, Dr. Smith needed to collaborate with a molecular biologist to obtain knockout mice for his behavioral experiments because the techniques for generating these mice were expensive and difficult to make work. In contrast, molecular biologists with transgenic mouse lines could buy an elevated plus maze and produce experimental results without ever speaking to a behaviorist. Asymmetries between behaviorists and molecular biologists were evident at the institutional level as well. While funding from the National Institute of Mental Health for genetic studies in animal models increased in the late 1990s, most of this investment went into knockout mouse models and large-scale mouse mutagenesis projects (Hyman 2006), and not the selective breeding projects preferred by researchers at Coast.

The intense concern with methodology evident among Coast practitioners, then, may reflect their relatively weak position in many of the scientific fields in which they participated. In his history of psychological research (another field in which some Coast researchers claimed membership), Kurt Danziger (1994, 5) argues that psychologists' anxiety about the status of their discipline led to a kind of methods fetishism or, in his words, "methodolatry." Danziger argues that debates about method served as a surrogate for more difficult debates about the value of psychological research. Moreover, by adopting methods based in quantification, he argues that psychologists were able to make alliances with powerful partners such as established scientific fields, industry, and the military. Researchers at Coast likewise emphasized the importance of experimental controls to combat perceptions of behavioral research as an overly "soft" science, and to exert authority over the interpretation of behavioral experiments in an environment where they were unable to control the uptake of these techniques into new disciplines.

A second factor contributing to the unique culture of claims making at Coast was the specific kind of modeling work researchers were engaged in. Many existing studies of model organism work have focused on places where researchers were using experimentally tractable organisms to elucidate general biological principles. Scientists' use of the fruit fly *Drosophila* as a model for investigating the principles of inheritance (Kohler 1994) or the nematode *C. elegans* as a model for understanding the nervous system (Ankeny 2001; see also de Chadarevian 1998) are both canonical examples of this approach. Rachel Ankeny (2007) describes *C. elegans* researchers' approach to knowledge production as akin to "case-based reasoning" in medicine, in that researchers used comparisons between individual organisms to make arguments about developmental processes writ large. Carrie Friese and Adele Clarke (2012) show how arguments about generalizability enable many aspects of modeling work. The argument that some animals can be treated as "more or less the same" as other species (42), for example, facilitated the substitution of cow eggs for human eggs in stem cell projects or of eggs from domestic animals for those of endangered species in zoo cloning projects (see also Friese 2009, 2013).

Researchers at Coast, in contrast, were not aiming to develop general theories of addiction that would apply to an entire class of organisms. Graduate student Alex joked to me that his lab was not looking to "solve the rat cocaine problem," and Dr. Smith told me more plainly that he was "not so interested" in uncovering universal biological truths. Rather, they had the more instrumental goal of using animal experiments to generate findings that would help address the problems of human addicts. To the extent that Dr. Smith was interested in cross-species comparisons, it was in trying to explain why humans seemed to have a unique susceptibility to particular drug addictions while other organisms did not. The research taking place at Coast did resemble other model organism work in that the rodents were "not being studied because they were interesting in their own right" (Ankeny and Leonelli 2011, 321), but Coast researchers' epistemological aim was not to generalize about behavior—they aimed to move not from the specific to the general but from the specific to the specific.

This may seem a subtle difference, but it is a significant one, because the arguments and evidence required to establish the mouse as a representative of a larger class of biological organisms are quite different from those required to establish a mouse pressing a lever for alcohol as a convincing analog for human addiction. Some justifications Coast researchers used for behavioral animal models (such as the ethological argument for the elevated plus maze) relied on evidence of evolutionary conservation between mice and humans, but many did not. Pharmacological arguments for behavioral tests were agnostic on the issue of whether animals and humans shared conserved biological or mental properties; they claimed only that animal experiments were useful tools for making predictions about humans. Animal models of addiction relied on evidence that mice and humans shared similar genes and brain structures, but practitioners were quick to point out that rodents did not share humans' biological susceptibility to alcohol addiction. Attempts to claim that anxiety or addiction were common to higher mammals could actually detract from the perceived validity of anxiety models in the eyes of some researchers, because they opened up contentious questions about animals' mental experiences. Developing psychiatric animal models, then, was not a matter of working out behavioral principles in mice and extrapolating them out to other mammals. Rather, it was a process of "unmaking" a mouse as a representative of its own species, and then "remaking" it into something more like a human addict (Svendsen and Koch 2013).

The specific epistemological aims of the modeling work taking place at Coast might explain why practitioners were as likely to push down on their scientific claims as they were to build them up. Their work did not depend on or aim for generalizations across species to the same extent that other kinds of model organism work does, and so they were not incentivized to enlarge their claims as much as possible. Broad generalizations could even be actively harmful to their research program: in a situation where the artifactual nature of a mouse's behavior in a drinking experiment was something that practitioners acknowledged and even embraced, generalizations became risky.

A final force contributing to the unique culture at Coast was its temporal location in what has come to be called the postgenomic era. Numerous analysts have described the complexity talk that practitioners at Coast engaged in as something characteristic of science after the completion of the Human Genome Project (HGP). Awash in genomic data showing evidence of alternative splicing, epigenetic imprinting, and transcriptional elements, genome scientists are increasingly talking about biological complexity. Hallam Stevens and Sarah Richardson (2015) argue that genome scientists today narrate their histories by tracing out a shift from simplicity to complexity, with the HGP as the turning point in these stories. Stevens and Richardson point to the headline of a news feature in *Nature*'s "Genome at Ten" issue—"Life Is Complicated"—as evidence of this narrative arc: the feature article begins with the assertion that biology was once considered a "simple science," but argues that the flood of data generated by the HGP has undermined these assumptions and forced a recognition of the genome's complexity (Hayden 2010, 664). Evelyn Fox Keller (2015, 16–17) similarly argues that over the course of the HGP, "our view of the genome as simply a collection of genes has gradually given way to a growing appreciation of the dynamic complexity of the genome's architecture."[1] Talia Dan-Cohen (2016, 902) writes that scientists are now operating in the midst of a "flourishing complexity discourse" and finding new ways of situating themselves in line with or in opposition to this rhetoric.

Some analysts have described this emerging complexity discourse as a way for scientists to diffuse criticisms that the HGP has fallen far short of its promised outcomes. A *New York Times* article marking the ten-year anniversary of the project, for example, opened with the claim that "medicine has yet to see any large part of the promised benefits" a decade after the first draft of the sequence (Wade 2010). The article goes on to explain that the time line to new treatments will be longer than initially predicted because "it has become clear that the genetics of most diseases are more complex than anticipated" (Wade 2010). Such narratives about unrealized hopes and the problems of complexity are common in behavior genetics. Excitement about the molecular revolution of the 1990s quickly gave way to disappointment in the field as durable associations between particular genes and psychiatric disorders proved more difficult to make than anticipated. Failures to replicate associations such as a reported link between alcoholism and the gene DRD2 (Gelernter et al. 1991) prompted commentaries from behavior geneticists themselves on whether such gene association studies were "even worth doing" (Kidd 1993, 71).[2]

Practitioners hoped that new paradigms such as Avshalom Caspi and Terrie Moffitt's gene–environment interaction studies (Caspi et al. 2002; Caspi et al. 2003) might provide a pathway toward more stable results, but even these celebrated studies have faced replication troubles (Tabery 2014). In their studies of psychiatric genetics review articles published between 1999 and 2008, Arribas-Ayllon and colleagues (2010, 20) argue that practitioners began to adopt a "simple-to-complex" narrative about the history of the field to counteract criticisms about their lack of productivity. This narrative explained why practitioners had had relatively little success in identifying genes for particular psychiatric disorders. "Complexity" exonerated the field from its failures, insulated it from criticisms of genetic determinism, and constructed a careful sense of optimism about the potential of new methodologies and multidisciplinary collaborations. Brian Wynne (2005) points out, however, that contemporary genome scientists are subject to other opposing forces—the demands of industry, of regulatory science, the presumed necessity to simplify science for the public—that encourage reductionist ways of talking about genetics. He concludes that postgenomic science is Janus-faced: it speaks in self-contradictory terms, and appears to be "complexity-oriented" and reductionist at the same time (76).[3]

It is difficult to say whether the ubiquitous complexity talk at Coast was a product of the postgenomic era and its attendant criticisms and economic incentives. Panofsky (2015) points out that debates about the relative importance of genetic and environmental factors are not new for behavior genetics; in many ways these discussions have been present since the field's inception. At Coast, researchers described themselves as having long had an appreciation of complexity, as evidenced by their continued choices to study gene products as they were expressed in live, behaving animals rather than in tissue cultures or in vitro systems. But their own historical narratives also described how their expectations had changed over time, especially concerning the number of genes they expected would be involved in a psychological disorder and how much of a difference they believed each individual gene could make. Regardless of whether there has been a change in the frequency with which researchers at Coast used the term *complexity*, the changing linguistic landscape of the life sciences surely gave new weight to this talk. These broader conversations made "complexity" an especially powerful phrase to invoke when arguing for particular research practices or about the likely trajectory of their research efforts.

Coast researchers' preoccupation with methodology, cautious claiming, and complexity was to some extent unique even among their fellow behavior geneticists. But considering the forces that might have given rise to this particular epistemic culture points to ways in which elements of that culture might extend beyond this individual site. Animal behavior geneticists are not the only group of practitioners who have found themselves in weak positions vis-à-vis other fields, and responded by drawing on an idealized vision of careful science to compensate for their lack of resources or cultural capital. Nor are they the only researchers using animals to model specific processes rather than generalizable, cross-species phenomena. In many other areas of biomedical research—where researchers are using pigs to understand digestive tract problems in preterm infants (Svendsen and Koch 2013) or using techniques developed for studying tobacco mosaic virus as models for polio research (Creager 2001)—arguments for particular models are being made on more limited, pragmatic grounds. And researchers at Coast are part of a chorus of contemporary voices calling for a greater appreciation of complexity in the life sciences.

I have gestured at multiple studies throughout this book where the scientists under examination shared positions and practices in common with those at Coast. Gusterson's (2008) weapons scientists also wrestled with the artifactual nature of their weapons simulations. Some even reached a moment resembling the complexity crises I described in chapter 1, where they doubted whether they could produce the desired data using the tools they had. Greiffenhagen and Sharrock's (2011) mathematician interlocutor shared some of the concerns about public communication I described in chapter 6, such as worries that mathematics textbooks presented an image of the field that looked overly certain and might thereby mislead the public. Landecker's (2013) endocrinologist alternated between seeing animal environments as something to be controlled and something valuable to study in their own right, in a way that closely resembles the account I gave in chapter 4 of how Coast researchers managed their epistemic products and by-products. And like the members of the Alcohol Research Group I described in chapter 5, Edwards's (2010) climate scientists also saw benefit in allowing multiple weather forecasting models to proliferate in the hopes that the strengths and limitations of each individual model would balance out.

Many aspects of Coast's epistemic culture reflected broader cultural beliefs about what science is and how it should operate. I noted in chapter 2 that Coast researchers' cautious claiming practices resembled Merton's (1973 [1942]) idealized image of scientific practice, particularly his norms of disinterestedness and organized skepticism. Researchers at Coast engaged these long-standing beliefs about what science is (or what it ought to be) by encouraging methodological critique and celebrating moments where researchers were willing to "destroy [their] own data." The epistemological modesty on display at Coast is something that Haraway (1997, 24) has described as "one of the founding virtues of what we call modernity." She drew on Steven Shapin and Simon Schaeffer's (1989) account of Robert Boyle's experiments to describe the emergence of what she called the "self-invisible modest man"—a witness whose experimental reports could be considered credible because of his self-discipline and ethical restraint, and his unmarked male body. Only those who could "disappear modestly" into the laboratory's "culture of no culture" (25) could be a reliable witness, Haraway argued. Researchers at Coast worked within and against the trope of the self-invisible modest man. They held strongly to virtues of epistemological modesty, even as they noted how their bodies and their smells shaped the behavior of their mice. But as Despret (2004, 118) has argued, pointing out how humans and experimental animals interact may simply be a different way of performing self-invisibility—in taking careful note of how s/he might "contaminate" the experiment, the experimenter suggests that s/he is capable of eradicating or at least neutralizing these effects. In some ways, then, the face of laboratory science at Coast is a very familiar one.

My hope is that the specific epistemic culture I have described in this book will be both familiar and strange, highlighting elements of scientific culture that I think are widespread but undertheorized. In her philosophical discussion of complexity and science, Sandra Mitchell (2009) argues that there is a "mismatch" between philosophers' theories of science and contemporary scientific research because philosophical theories have been worked out in a limited set of scientific domains—namely, the physical sciences. By focusing on fields such as climate change and behavioral sciences that embrace complexity and deep uncertainty, she aims to "extend and revise [philosopher's] epistemological framework" (12). Genetics laboratories such as the one I have studied here are hardly understudied in STS, but the aspects of laboratory practice that are especially visible in places where scientists have embraced complexity deserve more attention. We have sophisticated means of talking about how scientists eliminate uncertainty from their claims or downplay the "tacit knowledge" that is essential for making experiments work, but our accounts of how scientists preserve the uncertainty and the secondary knowledge gained through laboratory work are less developed. Similarly, we have systematic accounts of how scientists generate individual facts, but our ways of talking about how scientists build up the foundations of their long-term research programs are more diffuse.

In describing the construction of epistemic scaffolds and the management of epistemic by-products, my aim is to expand our analytical vocabulary and make it easier to talk about the variety of facts and processes that occur in the laboratory. Existing descriptions of the construction of scientific facts lend themselves to a pointillist view of laboratory work—one where scientists produce discrete, tightly focused claims that together form an understanding of a subject. This is but one of many kinds of scientific claims and techniques for forming an image of the natural world. Like painters, scientists may spend time applying base layers of claims and evidence that provide support for other claims but are never intended to be visible in the final product. They may mask off areas of study such as the animal mind to discourage fellow researchers from laying claims there. Or they may apply a wash over whole areas of a field to change the tone of how existing claims are perceived. It is time to start using a more varied vocabulary to discuss these diverse processes and especially to stop painting them with broad brushes such as "reductionism." In an era where "complexity" is at the tip of so many tongues, these words will no longer do.

ACKNOWLEDGMENTS

I received numerous sources of funding over the years that have made this book possible. The Institute for the Social Sciences at Cornell University and the Syracuse University Mellon Humanities Project Grant both provided seed funding to conduct exploratory research for this project, and a National Science Foundation Doctoral Dissertation Improvement Grant (Award No. SES 0749635) and a Doctoral Fellowship by the Social Sciences and Humanities Research Council of Canada funded the fieldwork for this study. A Fall Competition grant from the Office of the Vice Chancellor for Research and Graduate Education (VCRGE) at the University of Wisconsin (UW)-Madison provided funding to collect and analyze the final pieces of data I needed to finish the project. Finally, support for the publication of this book was provided by the VCRGE at UW-Madison, with funding from the Wisconsin Alumni Research Foundation.

The scientists at "Coast University" may remain unidentified in this book, but they are very real people who contributed enormously to this project. Fieldwork can be an awkward process—especially so when one is first learning the ropes—and the scientists I worked with were remarkably patient and generous with me as I learned to articulate my research questions and found my footing as an ethnographer. I thank "Dr. Smith" in particular for welcoming me into his laboratory and for his willingness to take me and my project seriously even though the kind of research I was engaged in was so different from his own.

I have given many talks based on the material in this book, including numerous presentations at the Society for the Social Studies of Science and the History of Science Society; lectures at the University of Exeter's Egenis Center, York University's STS seminar series, MIT's STS program, and Northwestern's Science in Human Culture program; and papers at the "History of Skill in Science and Medicine" workshop at McGill University and the "Working across Species" workshop at King's College London. My thanks go to all those whose questions, comments, and invitations to present have helped me work out these arguments over the years. I would especially like to thank Sophia Roosth for her thoughtful commentary on an earlier version of chapter 5 and William Deringer for suggesting the pseudonym "Coast University" to me. Some of the arguments I make in this book have also been worked out in publications. Chapter 3 is based on an article first published in *Social Studies of Science* (Nelson 2013), and chapter 2 contains elements of an article published in *Medical History* (Nelson 2015).

I was honored to win a First Book Award from the Center for the Humanities at UW–Madison while working on this book. As part of this award, the center convened a seminar on my behalf where a fantastic group of senior scholars read and commented on an early version of the book: Katarzyna Beilin, Angela Creager, Joan Fujimura, Hugh Gusterson, Richard Keller, Gregg Mitman, Lynn Nyhart, and Robert Streiffer. The feedback I received in this seminar improved the book in many ways, perhaps most importantly by giving me the motivation I needed to go back in and make major revisions to the later chapters. I would also like to thank a few other scholars who were generous enough to read and comment on the whole manuscript: Jenni Lieberman, whose precision with language helped me avoid slippages between actors' and analysts' usages of "complex"; and the two anonymous University of Chicago Press readers, whose feedback was especially valuable for situating this book in relation to different literatures.

Two research assistants helped me at various stages of the project: Melissa Charenko and Brita Larson. Brita's natural gift for copyediting has greatly improved the readability of the book and my writing in general. If you find this book (relatively) easy to follow, thank Brita.

The majority of this book was drafted in coffee shops around Madison while sitting next to Robert Streiffer and other members of our informal Writing Accountability Group (WAG). The positive peer pressure that the group provided helped me establish a daily writing practice, and WAG transformed what used to feel like an isolating experience into a much more enjoyable shared practice. Fellow WAGgers, it has been a pleasure to watch all of your articles and books grow as I have been working away at mine.

I have benefitted from a number of wonderful mentors over the course of my career, including Alberto Cambrosio, Peter Keating (who helped me come up with the title for this book), and Michael Lynch. Three more deserve special mention here. My tenure mentor Florence Hsia has been instrumental in creating a space in which I could actually get writing done in my first few years on the tenure track. She has been fiercely protective of me through the normal travails of starting a new job and some extra bumps incurred through administrative restructurings. I'm certain she has done more work than I can even guess behind the scenes to ensure that my research could thrive.

Sergio Sismondo has been mentoring me unofficially for years as the journal editor who guided me through the process of publishing my first article from this project. I asked him to be my external mentor because I was sure that after sending my manuscript back to me so many times for revisions that he would have no qualms telling me if I was messing up. What I did not anticipate, though, is that he would also become an important source of support and encouragement. Sergio read chapters of this book and provided feedback at key moments when my confidence in the work was faltering. In this and other ways, he has kept me from ruminating and helped me keep moving forward.

Steve Hilgartner has been with me and with this project from the dissertation stage, and he has continued to read drafts long after my graduation. The further I advance in my career and the more clearly I see how many other responsibilities were competing for his time and addition, the more I have come to appreciate what a gift it is that he has continued to make time for me. Steve has shaped my thinking in lots of ways—I've joked to other people that after so many years of working with him, I can hear him saying, "Dynamic tensions!," in my head whenever I start to settle too far into one way of seeing the world. But what has stayed with me most is the way he made academia feel like a process of collective problem solving rather than a series of critiques and counterarguments. That is undoubtedly one of the reasons that the conversations I have had with Steve have been some of the most enjoyable of my academic life.

Last, some thanks to my family. My parents, Geoff and Judy, reminded me that I would be remiss in publishing a book about genetics without thanking them for my genes, but as we know from twin studies it's hard to separate genetic effects from those of supportive early environments—so my thanks to them for both. My brother, Dan, and especially my sister, Laura, have always been absolutely confident that I would finish this book and it would be great.

There is nothing like the faith of a little sister to make you feel like you can do difficult things, or like the security of knowing that she would not have loved me a whit less had I never finished at all. And finally, my deepest thanks go to my partner Harald Kliems. Predictably, many things in my life started to slide in periods where I was deep into writing. Harald was always nearby, quietly shoring things up. Harald, I look forward to enjoying the postbook life with you.

APPENDIX

METHODS

Interviews

All interviews were conducted in a semistructured, conversational fashion. I asked a few standardized questions of all interviewees (e.g., "Tell me the five things you consider to be most important to control for in your experimental work"), but in most cases, the wording of the questions, the order of the questions, and the topics covered varied in each interview. I used the same interview guide for all interviews with graduate students: I first asked how they became interested in a career in research; about their first experiences in the laboratory; who taught them the techniques that they used and what kinds of things were emphasized to them during training; and what techniques they found easy or difficult to learn. For students who had worked in multiple laboratories or had completed their first-year rotations through several laboratories at Coast, I asked them to compare between the laboratories they worked in. I also asked graduate students about what they thought about the validity of the animal models that they were using; how they envisioned their research would be used in the future; and how they discussed their work with friends and family. For interviews with principal investigators and other actors such as veterinarians or staff in mouse labs, I asked questions based on the participant's particular area of research and involvement in methodological discussions. In some cases, I asked participants about their research histories, but in most cases, I used methodological issues arising from their own research as the starting point for interviews. I also asked these researchers questions similar to those asked of graduate students about model validity, future applications, and the public communication of behavior genetics research.

In total, I conducted interviews with fifty-two individuals between 2006 and 2009, and some individuals were interviewed multiple times. I recorded these with a digital audio recorder and conducted them in person whenever possible. Two interviews were conducted over the phone; five interviews were conducted in person but not recorded due to varying circumstances (e.g., the participant did not want to be recorded, or the interview took place in a venue where recording was not possible). For unrecorded interviews, I took rough notes during the interview and wrote down detailed notes about the interview within one day of the interview (and usually immediately after the interview). The interviews lasted between one and three hours. The table below provides a complete list of all of the individuals interviewed in this project, a general description of their position and location, the date of the recorded interview (if applicable), and the pseudonym that they are identified with in the text (if applicable). The table also lists the names and positions of individuals I did not interview but who appear in my field notes.

News Media Articles

A research assistant, Brita Larson, compiled a database of news media coverage of mouse research on anxiety and alcoholism using the LexisNexis database. We limited our search to selected major North American daily newspapers (*Washington Post*, *New York Times*, *Globe & Mail*, and *National Post*) to create a data set of a manageable size, and we limited our date ranges from January 1, 1985 (the first year LexisNexis had coverage for all of the newspapers selected) to December 31, 2012 (extending a few years beyond my fieldwork). We used the search string "(mice OR mouse) AND research AND (anxiety OR alcoholism)," in combination with a series of "AND NOT" terms to exclude unrelated articles on topics such as computer input technologies or Walt Disney characters. Larson checked each "AND NOT" term as it was added to the search string to ensure that the term excluded only articles that did not discuss biomedical research. The final list of exclusion terms was "Mickey Mouse," "Logitech," "spies," and "bush." Finally, Larson manually cleaned the resulting set of articles to further remove unrelated results (such as a *Washington Post* article describing parental anxiety about mouse droppings found in a daycare center). The final data set included a total of 105 newspaper articles.

In addition to this set of articles, another research assistant, Melissa Charenko, compiled a collection of news media coverage of specific animal behavior genetics articles in major US and world publications. For example, Charenko searched for articles covering the multisited mouse study (Crabbe, Wahlsten, and Dudek 1999), using search strings that included the names of the study authors (e.g., "John Crabbe") and keywords ("lab environment AND mice AND behavior"), delimited to the years between the publication of the study and 2015.

Data Analysis

To facilitate data analysis, I transcribed all recorded interviews in full; and coded all interview transcripts, field notes, and newspaper articles using an approach informed by grounded theory (Glaser and Strauss 1967). For the field notes and interviews, I started by coding a sample of my interview transcripts, freely generating descriptive codes that reflected the topics, terms, concepts, and problems that were present in the interviews. This process of open coding generated a list of about 150 codes. From this list, I developed a focused list of codes, which I used to code the entire set of transcripts and field notes. Larson and I used a similar approach to coding the newspaper article data set, first developing a set of descriptive codes through an open coding process, developing a focused set of codes, and then recoding the entire data set. I also used the "autocode" features of the qualitative data analysis software programs to search for and code specific terms in the data sets (I used ATLAS.ti for field notes and interview transcripts and NVivo for the newspaper articles). For example, I used autocoding to identify instances where researchers used permutations of the terms *complexity* or *complex* to describe some aspect of their work and to identify where the phrase *gene for* appeared in newspaper articles.

During the coding process and in analyzing the resulting collections of quotes, I kept in mind the grounded theory principle of "constant comparison," which encourages the analyst to examine the similarities and differences between the new information s/he tags with a particular code and the information s/he has already collected up under that code. I looked for consistencies in how researchers talked about a particular topic (such as factors to control for that were mentioned over and over), and differences in how the same topic was discussed in varying circumstances (such as the ways that researchers control mouse genomes versus mouse housing, the laboratory test environment, or their own bodies) and for the most divergent opinions on a particular topic (such as extremely conservative or permissive attitudes toward controls). I also employed some of the "cartographic" approaches to data analysis described by Adele Clarke (2005). Clarke argues that one of the limitations of grounded theory is that in focusing on the concerns raised by actors, it does not offer a straightforward route to uncovering silences and absences in the data. She offers several techniques as supplements to grounded theory that allow researchers to investigate the "situations" in which action takes place. In particular, I employed Clarke's technique of "positional mapping" to lay out the major arguments made and positions taken in particular debates or controversies. The technique of positional mapping is useful for visualizing the similarities and differences between the arguments and positions that actors take, as well as for identifying arguments or positions that are not taken (e.g., the notable absence of the argument in my data set that mouse anxiety is the same as human anxiety).

A NOTE ON ANONYMITY

Given the history of public controversy surrounding animal behavior genetics research and concerns about animal rights activists in the area around Coast University, I eventually chose to report on my fieldwork and interviews almost entirely anonymously. While many researchers (such as Dr. Smith) were willing to have their real names used—and some even encouraged me to use their real names to combat the culture of silence they felt surrounded animal research in their local area—it was obvious from the very beginning of my fieldwork that others were suspicious of my presence and what I might say about their research. While several researchers voiced concerns that I might be an undercover animal rights activist, other theories that they jokingly put forward to explain my presence at Coast were that I was an organizational sociologist who was comparing laboratories that functioned well with laboratories that were badly run, or that I was collecting gossip to conduct a study of scientific hookup culture. The idea that a nonscientist would be so interested in the mundane details of their protocols simply did not seem plausible to many researchers. In such an atmosphere, the consent forms that I produced at the beginning of every interview and the ensuing discussion about whether participants preferred to have their quotes used anonymously were helpful in developing trust. Researchers who had experience working with human genetic material were used to the requirement that they work only with "de-identified" samples, and so offering a promise of anonymity made my research techniques look more familiar and legitimate to them.

While conducting my interviews, I offered participants the option of being identified or having their interview material used anonymously. However, as I began to write, I realized that I could not identify some individuals in the text while providing an adequate level of anonymity to others. Principal investigators, for example, were often happy to be interviewed "on the record," but their graduate students and technicians were not. Unlike their supervisors, they had little experience giving interviews and were afraid of saying something "wrong." They were especially uncomfortable with talking openly about differences between laboratories or differences between their interpretations and those of their principal investigator, lest this information compromise their credibility in the eyes of their supervisors or potential future employers. I therefore chose to use pseudonyms for all of the individuals I interacted with at Coast University to mitigate against the possibility of identifying individuals who wanted to remain anonymous by association with named individuals. The names of consortiums and some protocols have also been replaced by generic pseudonyms (with the exception of the Mouse Phenome Project at the Jackson Laboratory, which was relatively independent from the other sites I studied and therefore did not raise the same problems of compromising anonymity through association). In some instances, such as my discussion of the research conducted by the "Alcohol Research Group," I chose to omit citations to published and unpublished documents from this group entirely to avoid creating compromising links between real names and pseudonyms (I use quotation marks to distinguish specific phrases contained in these documents from my own paraphrasing). Knowledgeable readers may still be able to identify this group or make educated guesses about the location of my field sites, but consistently using pseudonyms makes it much more difficult to carry these associations all the way back to the students and technicians who were most concerned with remaining anonymous.

The ethnographic convention of using pseudonyms is complicated by the fact that scientists are also in some sense public figures whose names appear in publicly available documents such as publications, press releases, or news articles. I chose to use a mixture of real names and pseudonyms to address this problem. When citing interviews or field notes, I use pseudonyms; but when citing scientific publications or other publicly available documents, I use the real names of the authors. This means that in a few cases an individual may be identified both by their real name and by a pseudonym at different places in the text. To reduce some of the confusion about which names are real and which are pseudonyms, I refer to individuals as "Dr." followed by their last names (or by their first names only in the case of graduate students and technicians) when using pseudonyms, and I omit the "Dr." when using real names. The list of informants and interviews below is also intended as a quick reference for readers who are confused about whether a particular name is a pseudonym or a real name.

List of Interviews and Informants

Position	Pseudonym (if applicable)	Interview date (if applicable)
Coast University		
behavior geneticist	Dr. Rachel Jackson	Sept. 2007
postdoc	Dr. Marcus Lam	
behavior geneticist	Dr. Laura Martin	March 2008, April 2008
behavior geneticist	Dr. Daniel Smith	Aug. 2006, Sept. 2007, April 2008
behavioral neuroscientist	Dr. George Taylor	
behavior geneticist	Dr. Ruth Tremblay	Feb. 2008
behavioral neuroscientist	Dr. Sherry Trudeau	May 2008[†]
animal care staff	Aiden	May 2008
graduate student	Alex	April 2008
graduate student	Ava	May 2008
graduate student	Chloe	April 2008
graduate student	Emily	March 2008
media training specialist	Eric	May 2008[†]
graduate student	Hannah	March 2008
graduate student	Ian	March 2008
technician	James	
laboratory manager	Jeffrey	
technician	Kimberly	
graduate student	Liam	March 2008
technician	Madeline	
graduate student	Matthew	April 2008
technician	Susan	
behavior geneticist		Sept. 2007
behavioral neuroscientist		May 2008[†]
behavioral neuroscientist		May 2008[†]
graduate student		March 2008
graduate student		March 2008
information technology manager		Sept. 2007
postdoc		Sept. 2007
technician		Sept. 2007

(*continued*)

Position	Pseudonym (if applicable)	Interview date (if applicable)
Mouse Phenome Project		
MPP director	Molly Bogue*	Jan. 2008
genetic resources manager		Jan. 2008
geneticist		Jan. 2008
phenotyping center coordinator		Jan. 2008
project staff		Jan. 2008
Other United States		
behavior geneticist	Dr. Linda Anderson	Nov. 2009
behavior geneticist	Dr. Scott Clark	Aug. 2006
behavior geneticist	Dr. David James	April 2008
veterinarian	Dr. Amy Lee	Oct. 2008
behavior geneticist	Dr. Charles Westin	Jan. 2009
behavior geneticist	Dr. Frank White	June 2008
NIAAA official	Dr. Raymond Williams	June 2009
behavior geneticist	Dr. Larry Wilson	Dec. 2008
behavior geneticist		Nov. 2007
behavioral neuroscientist		Dec. 2008
graduate student		Dec. 2008
NIDA official		June 2009
phenotyping center coordinator		June 2009
phenotyping center staff		June 2009
phenotyping center staff		June 2009
phenotyping center staff		June 2009
technician		Dec. 2008
Canada		
behavior geneticist	Dr. Steve Fortin	Jan. 2009
behavior geneticist	Dr. Anthony Roy	July 2006
behavior geneticist		March 2009
graduate student		July 2006
Germany		
behavior geneticist	Dr. Thomas Schmidt	Aug. 2009
ethologist	Hanno Würbel*	Aug. 2007
behavior geneticist		Aug. 2009

Names marked with * are real names; interview dates marked with † were not audio-recorded.

NOTES

INTRODUCTION

1. See also Galison (1987) on the means by which researchers confront methodological difficulties and bring experiments to an end.

CHAPTER 1

1. Dr. Smith pointed out to me later that in fact the number of possible combinations was closer to two hundred, joking to me that he must have a tendency to "overestimate complexity."

2. Many of these issues were central to debates that took place around the publication of the DSM-5, and the way that the entries related to alcohol changed with the DSM-5's publication in 2013 illustrates animal researchers' concerns about working in a continually shifting landscape. The DSM-5 did away with the two separate categories that had existed in the DSM-IV (alcohol abuse and alcohol dependence) and replaced them with a single category called "alcohol use disorders." In addition to reshuffling existing criteria into new categories, the DSM-5 also added "craving" as a new criterion for alcohol use disorder.

3. Gail Davies (2013) has argued that generally speaking, such instances of "biological emergence" or "lively exuberance" on the part of mice are only rarely seen as valuable because they are disruptive to experimental research.

4. A "holding cage" is a clean cage that researchers use to temporarily house mice after they are taken out of their home cages but before they are placed in the experimental apparatus.

5. See Würbel (2002) for an example of the dozens of parameters that some behavioral researchers collect data on in their laboratories, including whether animal handlers were certified, smoked, or used perfume.

6. See, for example, Rosenthal and Fode (1963), where they showed that telling experimenters that particular rats were "bright" or "dull" impacted the speed with which rats learned to navigate a maze.

7. The reader, at this point, might be wondering the same thing!

CHAPTER 2

1. See Silva's autobiographical note on his personal website, http://www.silvalab.org/alcino_silva.html (accessed January 26, 2015).

2. Two of the four knockout mouse lines used in Kandel's 1992 paper were developed in Philippe Soriano's laboratory at Baylor University, which Soriano likely provided to Kandel in exchange for coauthorship on the paper. For more on exchange practices within the mouse community, see Murray (2010).

3. Jacqueline Crawley, personal communication, November 21, 2007.

4. Diane Paul (1998) likewise argues for the centrality of animal research in the emergence of the behavior genetics field post–World War II.

5. See also Hedgecoe and Martin (2003) on the importance of future visions in constructing sociotechnical networks.

6. The first of several books critiquing the publication was titled *The Bell Curve Debate* (Jacoby and Glauberman 1995) and collected critical commentaries from more than eighty different authors into a volume nearly equal in length to *The Bell Curve* itself.

CHAPTER 3

1. For an example of the "gold standard" language, see Crawley (2007, 262).

2. Other scholars in the history and philosophy of science have also employed scaffold metaphors, although for different purposes than I do here. Linnda Caporael, James Griesemer, and William Wimsatt (2013) have used the term *scaffolding* to describe how phenomena such as cognition or culture emerge over time (see also Wimsatt and Griesemer 2007). They borrow the term from developmental psychology, where the scaffold metaphor describes how parents help their children acquire difficult skills by supporting the process of learning (e.g., by creating a safe environment in which to practice dangerous tasks or creating approximations of more complex tasks). Their aim is to contribute to theoretical biology by providing a framework for understanding the evolution of complex phenomena, while my interest is more sociological: I want to understand how researchers make claims about the capacities of their scientific tools and how those claims are negotiated in scientific communities.

3. The association of ethology with a rejection of laboratory studies and an embrace of field research may make this seem an odd choice of words. I have chosen to describe these arguments as "ethological" primarily because researchers themselves use this term, but it is also worth noting that ethology and laboratory studies are not so isolated as analysts often assume. Robert Kirk's (2009) work on Michael Robin Alexander Chance (a British ethologist who spent part of his career working in a pharmaceutical company) shows how some practitioners used ethological principles as a means to improve animal housing and experimental design in the laboratory. Kirk (2009, 533) goes so far as to argue that "ethology was the principle [*sic*] vector through which scientific and moral necessity came to be integrated within the material practice in the laboratory."

4. Researchers often use ethological information to explain variation in test results between laboratories, an example of the dynamic combination of different kinds of arguments that I

address in more detail below. This survey of laboratories using the elevated plus maze, for example, argued that the variation in the light levels in researchers' testing rooms could alter the animal's assessment of the potential danger of the open spaces of the maze and thereby account for variation in the test results (Hogg 1996).

5. As Myers (2015) and Vertesi (2015) have argued, the ways that scientists embody their research subjects can be both a means of performing modeling work and building a sense of community, both of which are evident in this brief interaction.

CHAPTER 4

1. He was referring to the agouti mouse model, where the mouse's coat color varies depending on nutritional and environmental differences that influence epigenetic changes established early in development. See, for example, Dolinoy (2008) for a description of how researchers use the mouse's change in coat color as a "biosensor" for environmental changes.

2. See Drake et al. (1998) for an example of a calculation of the rates of spontaneous mutations in inbred mouse strains.

3. Studies focused on care and welfare in the animal laboratory suggest that technicians' knowledge of the animals is often discounted (see, e.g., Birke, Arluke, and Michael 2007).

4. A holding cage is a clean, empty cage that researchers use to temporarily house mice after they have taken them out of their home cages. Returning a mouse to its home cage after doing a procedure (such as weighing it) was considered bad practice at Coast because the agitated mouse might impact the behavior of its cage mates on being returned to the home cage.

5. The term *tacit knowledge* is often used with much less specificity in the STS literature, blurring Collins's distinction between what is difficult to say and what is left unsaid.

6. Alberto Cambrosio and Peter Keating (1988) took issue with what they saw as an implicit assumption in STS work that scientific actors did not recognize tacit knowledge as important to their own work. They argued there is substantial overlap between descriptions of tacit knowledge offered by analysts and scientists' descriptions of the "art" or "magic" involved in getting particular techniques to work.

7. Lynch (1988), for example, argues that skills in and observations about handling laboratory rats are rarely acknowledged by the scientists as knowledge and so constitute a kind of "subjugated knowledge" that is masked by the dominant systems of knowledge production in the laboratory.

8. Sonic metaphors can be similarly problematic. While a squeaky mouse might bring to mind the persistence of the material world, other examples (such as signal and noise) allow for the interpretation that some features of the laboratory environment can be filtered out and forgotten.

9. See, for example, Würbel (2001) and Wolfer et al. (2004).

10. Robert Proctor and Londa Schiebinger (Proctor 1995; Proctor and Schiebinger 2008) coined the term *agnotology* to describe their studies of the resources and effort devoted to the creation of ignorance around particular issues, such as the link between tobacco use and cancer. Scott Frickel and colleagues (2010) have issued a similar call for attention to "undone" science, which they define broadly as research that is incomplete, ignored, or un(der)funded.

Paul Wenzel Geissler (2013) has used the term *unknowing* to draw attention to the active processes by which particular kinds of knowledge are suppressed.

11. See, for example, Gail Davies (2013) on how researchers attend to the "biological exuberance" of laboratory mice, or Viciane Despret (2004) on animal agency in the laboratory and beyond.

12. The reasons why some researchers might value what others regard as a nuisance are multiple: disciplinary training and affiliations surely matter, as do individual researchers' personal experiences or social position, funding incentives, commercial value of particular results, or styles of thought. Evelyn Fox Keller has explored many of these factors in her work, showing how gender and disciplinary traditions might explain why some researchers pursued observations and lines of inquiry that others ignored (1983, 1997), or how reductionist styles of thought might account for biologists' intense focus on genetics throughout the twentieth century (2000).

13. William Cronon's (1991) classic book *Nature's Metropolis* makes this point quite clearly. Thanks to Ken Alder for making this connection for me.

14. Würbel's publication record nicely demonstrates that what is tacit in some contexts might be published and widely shared in others. Far from being excluded from publication, his work on animal housing has appeared in prestigious venues such as *Trends in Neurosciences*, *Behavioral Brain Research*, and even *Nature*.

15. See Nelson (2015) for more on debates around knockout techniques.

16. Rosenthal has received, for example, a prize for his research from the American Association for the Advancement of Science and a lifetime achievement award from the American Psychological Association.

17. See, for example, Mansoor Niaz's (2015) description of how Robert Millikan discarded more than half of his experimental observations in his Nobel Prize–winning experiments calculating the value of the elementary electronic charge.

CHAPTER 5

1. Although making these measurements sounds quite straightforward, the specific design of the protocol made it possible to obtain blood alcohol levels with an ease that had not been possible before. Even if mice had been drinking large amounts in experimental setups such as the two bottle choice model, it would have been challenging for researchers to show evidence of this, because alcohol was available to the mice for large periods of time. Researchers had to guess at when the mice were drinking and when they should therefore take blood samples, and even if they made informed guesses or took multiple measurements, they might still catch only a few mice with high BACs at any given time.

2. The Massachusetts-based biotech company Biomodels, for example, uses the model as part of its preclinical pharmaceutical evaluation services.

3. Later work has offered more specificity on the origin and nature of the "disease concept" of alcoholism. Vrecko (2010a), for example, identifies the late 1960s and early 1970s as the key moment for the birth of the currently accepted conception of alcoholism as a "chronic, relapsing brain disease" (NIDA 2007)—a conception that analysts have dubbed the "NIDA

paradigm" to highlight the institute's role in securing this definition (Dunbar, Kushner, and Vrecko 2010).

4. Conrad and Schneider (1980), Room (1974, 1983), and Schneider (1978) were among the first to advance this argument, but variants of it have been put forward by many other scholars.

5. Again, Conrad and Weinberg (1996), Conrad (1999a), Lippman (1992), and Nelkin and Lindee (1995) were among the first to apply these critiques to genetics, but this general argument has been taken up widely by later scholars.

6. For more on Cicero's criteria, see Ankeny et al. (2014) and Ramsden (2015).

7. The extensive discussions of "motivation" within the ARG may seem out of step with the prohibitions against anthropomorphism and against speculations on the animal mind that I discussed in chapter 3. Here, motivation is generally used in a behaviorist sense to refer to an observable action (such as a willingness to "work for" access to alcohol) and not a mental state.

8. Other members of the group pointed out that there were also practical reasons to prefer one model over the other—namely, that the nocturnal drinking protocol could be completed in five days whereas the limited water model took up to twenty-one days.

9. Jamie Lewis and colleagues (2013) make the similar point that developing models that are "good enough" for particular practical purposes can also lead to a proliferation of models that each capture a different subset of a disorder.

CHAPTER 6

1. Nelkin and Lindee's 1995 book is one prominent example of this line of critique, but see also Duster (1990) and Alper and Beckwith (1993) for other similar arguments about behavior genetics.

2. While there is some research examining the degree of similarity between science journalists' and geneticists' views and values (e.g., Geller et al. 2005), there is little that specifically examines whether journalists as a group are more likely to hold determinist or reductionist views. Science reporter Deborah Blum's (1999) account of her experiences covering behavior genetics research offers some anecdotal evidence that journalists' views on gene action are similar to those of many behavior geneticists.

3. When the university had advance notice that a protest was being organized, they sent out an e-mail to university employees letting them know that they should prepare for possible interruptions and delays.

4. The image of laboratory animals as helpers or saviors is common in biomedical research, as Birke (2003) shows.

5. Numerous other science communication scholars have made similar points about news reports' emphasis on products over process, including LaFolette (1990) and Nelkin (1987).

6. Ironically the author, Jonah Lehrer, was playing fast and loose with the truth in his own writing: he was later found to have fabricated quotes for his publications and subsequently resigned from the *New Yorker*.

7. John Crabbe's letter to the editor does not appear to have been published, but an excerpt from it appears in a blog post on the *National Geographic* website.

CONCLUSION

1. Richardson (2015) offers a nice summary of other secondary literature that argues for a shift from gene-centrism to holism and antireductionism in the postgenomic era.

2. See Arribas-Ayllon et al. (2010) and Panofsky (2015) for more examples of failures to replicate in behavior genetics in the 1990s.

3. I find Wynne's description of the Janus-faced nature of postgenomics compelling, even though I disagree with the implicit assumption elsewhere in his argument that it is possible to judge the degree of reductionism inherent in scientific representations or scientific practice.

BIBLIOGRAPHY

Abbott, Andrew. 1988. *The System of Professions: An Essay on the Division of Expert Labor*. Chicago: University of Chicago Press.

Abir-Am, Pnina G. 1992. "The Politics of Macromolecules: Molecular Biologists, Biochemists, and Rhetoric." *Osiris* 7: 164–91.

Acker, Caroline Jean. 2010. "How Crack Found a Niche in the American Ghetto: The Historical Epidemiology of Drug-Related Harm." *BioSocieties* 5 (1): 70–88. doi:10.1057/biosoc .2009.1.

Alper, Joseph S., and Jon Beckwith. 1993. "Genetic Fatalism and Social Policy: The Implications of Behavior Genetics Research." *Yale Journal of Biology and Medicine* 66 (6): 511–24.

Anderson, Philip Warren. 1972. "More Is Different." *Science* 177 (4047): 393–96. doi:10.1126/ science.177.4047.393.

Ankeny, Rachel A. 2001. "The Natural History of *Caenorhabditis elegans* Research." *Nature Review Genetics* 2 (6): 474–79. doi:10.1038/35076538.

———. 2007. "Wormy Logic: Model Organisms as Case Based Reasoning." In *Science without Laws: Model Systems, Cases, Exemplary Narratives*, edited by Angela N. H. Creager, Elizabeth Lunbeck, and M. Norton Wise, 46–58. Durham, NC: Duke University Press.

Ankeny, Rachel A., and Sabina Leonelli. 2011. "What's So Special about Model Organisms?" *Studies in History and Philosophy of Science Part A* 42 (2): 313–23. doi:10.1016/j.shpsa .2010.11.039.

Ankeny, Rachel A., Sabina Leonelli, Nicole C. Nelson, and Edmund Ramsden. 2014. "Making Organisms Model Human Behavior: Situated Models in North-American Alcohol Research, since 1950." *Science in Context* 27 (3): 485–509. doi:10.1017/ S0269889714000155.

Arribas-Ayllon, Michael, Andrew Bartlett, and Katie Featherstone. 2010. "Complexity and Accountability: The Witches' Brew of Psychiatric Genetics." *Social Studies of Science* 40 (4): 499–524. doi:10.1177/0306312710363511.

Barad, Karen. 2003. "Posthumanist Performativity: Toward an Understanding of How Matter Comes to Matter." *Signs* 28 (3): 801–31. doi:10.1086/345321.

———. 2007. *Meeting the Universe Halfway: Quantum Physics and the Entanglement of Matter and Meaning*. Durham, NC: Duke University Press.

Berger, John. 1980. *About Looking*. New York: Pantheon Books.

Birke, Lynda. 2003. "Who—or What—Are the Rats (and Mice) in the Laboratory." *Society and Animals* 11 (3): 207–24. doi:10.1163/156853003322773023.

Birke, Lynda, Arnold Arluke, and Mike Michael. 2007. *The Sacrifice: How Scientific Experiments Transform Animals and People*. West Lafayette, IN: Purdue University Press.

Blum, Deborah. 1999. "Reporting on the Changing Science of Human Behavior." In *Communicating Uncertainty: Media Coverage of New and Controversial Science*, edited by Sharon Friedman, Sharon Dunwoody, and Carol Rogers, 155–66. Mahwah, NJ: Erlbaum.

Bogue, Molly A., and Stephen C. Grubb. 2004. "The Mouse Phenome Project." *Genetica* 122 (1): 71–74. doi:10.1007/s10709-004-1438-4.

Bourin, Michel, Benoit Petit-Demoulière, Brid Nic Dhonnchadha, and Martine Hascöet. 2007. "Animal Models of Anxiety in Mice." *Fundamental & Clinical Pharmacology* 21 (6): 567–74. doi:10.1111/j.1472-8206.2007.00526.x.

"Brain Link to Taste for Drink." 1998. *Courier Mail: News* [Queensland, Australia], December 5, 7.

Brown, Nik, and Mike Michael. 2003. "A Sociology of Expectations: Retrospecting Prospects and Prospecting Retrospects." *Technology Analysis & Strategic Management* 15 (1): 3. doi:10.1080/0953732032000046024.

Bubela, Tania M., and Timothy A. Caulfield. 2004. "Do the Print Media 'Hype' Genetic Research? A Comparison of Newspaper Stories and Peer-Reviewed Research Papers." *CMAJ* 170 (9): 1399–1407. doi:10.1503/cmaj.1030762.

Cambrosio, Alberto, and Peter Keating. 1988. "'Going Monoclonal': Art, Science, and Magic in the Day-to-Day Use of Hybridoma Technology." *Social Problems* 35 (3): 244–60. doi:10.2307/800621.

Campbell, Brian L. 1985. "Uncertainty as Symbolic Action in Disputes among Experts." *Social Studies of Science* 15 (3): 429–53. doi:10.1177/030631285015003002.

Campbell, Nancy D. 2007. *Discovering Addiction: The Science and Politics of Substance Abuse Research*. Ann Arbor: University of Michigan Press.

———. 2010. "Toward a Critical Neuroscience of 'Addiction.'" *BioSocieties* 5: 89–104. doi:10.1057/biosoc.2009.2.

Caporael, Linnda R., James R. Griesemer, and William C. Wimsatt, eds. 2013. *Developing Scaffolds in Evolution, Culture, and Cognition*. Cambridge, MA: MIT Press.

Carter, Adrian, Benjamin Capps, and Wayne D. Hall. 2012. "Emerging Neurobiological Treatments of Addiction: Ethical and Public Policy Considerations." In *Addiction Neuroethics*, edited by Adrian Carter, Wayne Hall, and Judy Illes, 95–113. San Diego, CA: Academic Press. doi:10.1016/B978-0-12-385973-0.00005-3.

Caspi, Avshalom, Joseph McClay, Terrie E. Moffitt, Jonathan Mill, Judy Martin, Ian W. Craig, Alan Taylor, and Richie Poulton. 2002. "Role of Genotype in the Cycle of Violence in Maltreated Children." *Science* 297 (5582): 851–54. doi:10.1126/science.1072290.

Caspi, Avshalom, Karen Sugden, Terrie E. Moffitt, Alan Taylor, Ian W Craig, HonaLee Harrington, Joseph McClay et al. 2003. "Influence of Life Stress on Depression: Moderation by a Polymorphism in the 5-Htt Gene." *Science* 301 (5631): 386–89. doi:10.1126/science.1083968.

Chesler, Elissa J., Sonya G. Wilson, William R. Lariviere, Sandra L. Rodriguez-Zas, and Jeffrey S. Mogil. 2002a. "Identification and Ranking of Genetic and Laboratory Environment Factors Influencing a Behavioral Trait, Thermal Nociception, via Computational Analysis of a Large Data Archive." *Neuroscience & Biobehavioral Reviews* 26 (8): 907–23. doi:10.1016/S0149-7634(02)00103-3.

———. 2002b. "Influences of Laboratory Environment on Behavior." *Nature Neuroscience* 5 (11): 1101–2. doi:10.1038/nn1102-1101.

Clarke, Adele. 2005. *Situational Analysis: Grounded Theory after the Postmodern Turn*. Thousand Oaks, CA: Sage.

Collins, Francis S. 2011. "Reengineering Translational Science: The Time Is Right." *Science Translational Medicine* 3: 90cm17. doi:10.1126/scitranslmed.3002747.

Collins, Harry M. 1974. "The TEA Set: Tacit Knowledge and Scientific Networks." *Science Studies* 4 (2): 165–85. doi:10.1177/030631277400400203.

———. 1985. *Changing Order: Replication and Induction in Scientific Practice*. London: Sage.

———. 2001. "Tacit Knowledge, Trust and the Q of Sapphire." *Social Studies of Science* 31 (1): 71–85. doi:10.1177/030631201031001004.

———. 2010. *Tacit and Explicit Knowledge*. Chicago: University of Chicago Press.

Collins, Nick. 2013. "Alcoholism Linked to Mutated DNA: 'Excessive Drinking' Gene May Play Key Role in Addictive Behaviour: Scientists." *National Post*, November 7. http://news.nationalpost.com/health/alcoholism-linked-to-mutated-dna-excessive-drinking-gene-may-play-key-role-in-addictive-behaviour-u-k-scientists-say.

Conrad, Peter. 1999a. "A Mirage of Genes." *Sociology of Health and Illness* 21 (2): 228–41. doi:10.1111/1467-9566.00151.

———. 1999b. "Uses of Expertise: Sources, Quotes, and Voice in the Reporting of Genetics in the News." *Public Understanding of Science* 8 (4): 285–302. doi:10.1088/0963-6625/8/4/302.

———. 2001. "Genetic Optimism: Framing Genes and Mental Illness in the News." *Culture, Medicine and Psychiatry* 25 (2): 225–47. doi:10.1023/A:1010690427114.

Conrad, Peter, and Joseph W. Schneider. 1980. *Deviance and Medicalization: From Badness to Sickness*. St. Louis, MO: Mosby.

Conrad, Peter, and Dana Weinberg. 1996. "Has the Gene for Alcoholism Been Discovered Three Times since 1980? A News Media Analysis." *Perspectives in Social Problems* 8: 3–24.

Coole, Diana, and Samantha Frost, eds. 2010. *New Materialisms: Ontology, Agency, and Politics*. Durham, NC: Duke University Press.

Crabbe, John C., Douglas Wahlsten, and Bruce C. Dudek. 1999. "Genetics of Mouse Behavior: Interactions with Laboratory Environment." *Science* 284 (5420): 1670–72. doi:10.1126/science.284.5420.1670.

Crawley, Jacqueline N. 2000. *What's Wrong with My Mouse? Behavioral Phenotyping of Transgenic and Knockout Mice*. New York: Wiley-Liss.

———. 2007. *What's Wrong with My Mouse? Behavioral Phenotyping of Transgenic and Knockout Mice*. 2nd ed. New York: Wiley-Liss.

Crawley, Jacqueline N., John K. Belknap, Allan Collins, John C. Crabbe, Wayne Frankel, Norman Henderson, Robert J. Hitzemann et al. 1997. "Behavioral Phenotypes of Inbred Mouse Strains: Implications and Recommendations for Molecular Studies." *Psychopharmacology* 132 (2): 107–24. doi:10.1007/s002130050327.

Creager, Angela N. H. 2001. *The Life of a Virus*. Chicago: University of Chicago Press.

Creager, Angela N. H., Elizabeth Lunbeck, and M. Norton Wise. 2007. "Introduction." In *Science without Laws: Model Systems, Cases, Exemplary Narratives*, edited by Angela N. H. Creager, Elizabeth Lunbeck, and M. Norton Wise, 1–22. Durham, NC: Duke University Press.

Crist, Eileen. 1999. *Images of Animals: Anthropomorphism and Animal Mind*. Philadelphia: Temple University Press.

Cronje, Ruth. 2008. "Going Public with the Scientific Process." *Science* 319 (5869): 1483. doi:10.1126/science.319.5869.1483d.

Cronon, William. 1991. *Nature's Metropolis: Chicago and the Great West*. New York: Norton.

Cwiartka, Monika. 2011. "How Do Mice Mean? The Rhetoric of Measurement in the Medical Laboratory." In *Rhetorical Questions of Health and Medicine*, edited by Joan Leach and Deborah Dysart-Gale, 33–58. Lanham, MD: Lexington Books.

Dan-Cohen, Talia. 2016. "Ignoring Complexity Epistemic Wagers and Knowledge Practices among Synthetic Biologists." *Science, Technology & Human Values* 41 (5): 899–921. doi:10.1177/0162243916650976.

Danziger, Kurt. 1994. *Constructing the Subject: Historical Origins of Psychological Research*. Cambridge: Cambridge University Press.

Daston, Lorraine, and Gregg Mitman, eds. 2005. *Thinking with Animals: New Perspectives on Anthropomorphism*. New York: Columbia University Press.

Davies, Gail. 2010. "Captivating Behaviour: Mouse Models, Experimental Genetics and Reductionist Returns in the Neurosciences." *Sociological Review* 58: 53–72. doi:10.1111/j.1467-954X.2010.01911.x.

———. 2013. "Mobilizing Experimental Life: Spaces of Becoming with Mutant Mice." *Theory, Culture & Society* 30 (7–8): 129–53. doi:10.1177/0263276413496285.

Dawson, Gerard R., Jonathan Flint, and Lawrence S. Wilkinson. 1999. "Testing the Genetics of Behavior in Mice." *Science* 285 (5436): 2068.

Dawson, Gerard R., and Mark D. Tricklebank. 1995. "Use of the Elevated Plus Maze in the Search for Novel Anxiolytic Agents." *Trends in Pharmacological Sciences* 16 (2): 33–36. doi:10.1016/S0165-6147(00)88973-7.

de Chadarevian, Soraya. 1998. "Of Worms and Programmes: *Caenorhabditis elegans* and the Study of Development." *Studies in History and Philosophy of Science Part C* 29 (1): 81–105. doi:10.1016/S1369-8486(98)00004-1.

Delamont, Sara, and Paul Atkinson. 2001. "Doctoring Uncertainty: Mastering Craft Knowledge." *Social Studies of Science* 31 (1): 87–107. doi:10.1177/030631201031001005.

Despret, Vinciane. 2004. "The Body We Care For: Figures of Anthropo-Zoo-Genesis." *Body & Society* 10 (2–3): 111–34. doi:10.1177/1357034X04042938.

Dietrich, Michael R., Rachel A. Ankeny, and Patrick M. Chen. 2014. "Publication Trends in Model Organism Research." *Genetics* 198 (3): 787–94. doi:10.1534/genetics.114.169714.

Dingel, Molly J., Jenny Ostergren, Jennifer B. McCormick, Rachel Hammer, and Barbara A. Koenig. 2014. "The Media and Behavioral Genetics Alternatives Coexisting with Addiction Genetics." *Science, Technology & Human Values* 40 (4): 459–86. doi:10.1177/0162243914558491.

"Diseases That Attack the Brain: Scientists Try a Novel Approach to Save Brain Cells." 1995. *Washington Post*, April 4, Z10.

Doing, Park. 2008. "Give Me a Laboratory and I Will Raise a Discipline: The Past, Present, and Future Politics of Laboratory Studies in STS." In *The Handbook of Science and Technology Studies*, 3rd ed., edited by Edward J. Hackett, Olga Amsterdamska, Michael E. Lynch, and Judy Wajcman, 279–95. Cambridge, MA: MIT Press.

Dole, Vincent P., and R. Thomas Gentry. 1984. "Toward an Analogue of Alcoholism in Mice: Scale Factors in the Model." *Proceedings of the National Academy of Sciences* 81 (11): 3543–46.

Dolinoy, Dana C. 2008. "The Agouti Mouse Model: An Epigenetic Biosensor for Nutritional and Environmental Alterations on the Fetal Epigenome." *Nutrition Reviews* 66 (suppl 1): S7–S11. doi:10.1111/j.1753-4887.2008.00056.x.

Drago, John, Charles R. Gerfen, Jean E. Lachowicz, Heinz Steiner, Tom R. Hollon, Paul E. Love, Guck T. Ooi, Alexander Grinberg, Eric J. Lee, and Sing Ping Huang. 1994. "Altered Striatal Function in a Mutant Mouse Lacking D1A Dopamine Receptors." *Proceedings of the National Academy of Sciences* 91 (26): 12564–68.

Drake, John W., Brian Charlesworth, Deborah Charlesworth, and James F. Crow. 1998. "Rates of Spontaneous Mutation." *Genetics* 148 (4): 1667–86.

Dunbar, Deanne, Howard I. Kushner, and Scott Vrecko. 2010. "Drugs, Addiction and Society." *BioSocieties* 5 (1): 2–7. doi:10.1057/biosoc.2009.10.

Dunwoody, Sharon. 1982. "A Question of Accuracy." *IEEE Transactions on Professional Communication* PC-25 (4): 196–99. doi:10.1109/TPC.1982.6447803.

———. 1999. "Scientists, Journalists, and the Meaning of Uncertainty." In *Communicating Uncertainty: Media Coverage of New and Controversial Science*, edited by Sharon Friedman, Sharon Dunwoody, and Carol Rogers, 59–79. Mahwah, NJ: Erlbaum.

Dussauge, Isabelle, Claes-Fredrik Helgesson, and Francis Lee, eds. 2015. *Value Practices in the Life Sciences and Medicine*. Oxford: Oxford University Press.

Duster, Troy. 1990. *Backdoor to Eugenics*. New York: Routledge.

———. 2003. *Backdoor to Eugenics*. 2nd ed. New York: Routledge.

Edwards, Paul N. 2010. *A Vast Machine: Computer Models, Climate Data, and the Politics of Global Warming*. Cambridge, MA: MIT Press.

Engber, Daniel. 2011. "The Trouble with Black-6." *Slate*, November. http://www.slate.com/articles/health_and_science/the_mouse_trap/2011/11/black_6_lab_mice_and_the_history_of_biomedical_research.html.

Enserink, Martin. 1999. "Fickle Mice Highlight Test Problems." *Science* 284 (5420): 1599–600. doi:10.1126/science.284.5420.1599a.

"Findings: Alcoholism and Neuropeptide Y." 1998. *Washington Post*, November 26, A05.

Fingarette, Herbert. 1988. *Heavy Drinking: The Myth of Alcoholism as a Disease*. Berkeley: University of California Press.

Flint, Jonathan, William Valdar, Sagiv Shifman, and Richard Mott. 2005. "Strategies for Mapping and Cloning Quantitative Trait Genes in Rodents." *Nature Reviews Genetics* 6 (4): 271–86. doi:10.1038/nrg1576.

Fortun, Michael. 1999. "Projecting Speed Genomics." In *The Practices of Human Genetics*, edited by Michael Fortun and Everett Mendelsohn, 25–48. Dordrecht: Springer Netherlands.

———. 2001. "Mediated Speculations in the Genomics Futures Markets." *New Genetics and Society* 20 (2): 139–56. doi:10.1080/14636770124557.

———. 2008. *Promising Genomics: Iceland and DeCODE Genetics in a World of Speculation*. Berkeley: University of California Press.

Fraser, Leanne M., Richard E. Brown, Ahmed Hussin, Mara Fontana, Ashley Whittaker, Timothy P. O'Leary, Lauren Lederle, Andrew Holmes, and André Ramos. 2010. "Measuring Anxiety- and Locomotion-Related Behaviours in Mice: A New Way of Using Old Tests." *Psychopharmacology* 211 (1): 99–112. doi:10.1007/s00213-010-1873-0.

Frickel, Scott, Sahra Gibbon, Jeff Howard, Joanna Kempner, Gwen Ottinger, and David J. Hess. 2010. "Undone Science: Charting Social Movement and Civil Society Challenges to Research Agenda Setting." *Science, Technology & Human Values* 35 (4): 444–73. doi:10.1177/0162243909345836.

Friedman, Richard A. 2015. "The Feel-Good Gene." *New York Times: Sunday Review*, March 8, 1.

Friese, Carrie. 2009. "Models of Cloning, Models for the Zoo: Rethinking the Sociological Significance of Cloned Animals." *BioSocieties* 4 (4): 367–90. doi:10.1017/S1745855209990275.

———. 2013. *Cloning Wild Life: Zoos, Captivity, and the Future of Endangered Animals*. New York: New York University Press.

Friese, Carrie, and Adele E. Clarke. 2012. "Transposing Bodies of Knowledge and Technique: Animal Models at Work in Reproductive Sciences." *Social Studies of Science* 42 (1): 31–52. doi:10.1177/0306312711429995.

Fujimura, Joan H. 1987. "Constructing 'Do-Able' Problems in Cancer Research: Articulating Alignment." *Social Studies of Science* 17 (2): 257–93. doi:10.1177/030631287017002003.

———. 2006. "Sex Genes: A Critical Sociomaterial Approach to the Politics and Molecular Genetics of Sex Determination." *Signs* 32 (1): 49–82. doi:10.1086/505612.

Galison, Peter. 1987. *How Experiments End*. Chicago: University of Chicago Press.

Garfinkel, Harold. 1956. "Conditions of Successful Degradation Ceremonies." *American Journal of Sociology* 61 (5): 420–24. doi:10.1086/221800.

Gartner Inc. 2016. "Research Methodologies: Gartner Hype Cycle." http://www.gartner.com/technology/research/methodologies/hype-cycle.jsp.

Gaudillière, Jean-Paul. 2001. "Making Mice and Other Devices: The Dynamics of Instrumentation in American Biomedical Research (1930–1960)." In *Instrumentation Between*

Science, State and Industry, edited by Bernward Joerges and Terry Shinn, 175–96. Dordrecht: Springer Netherlands.

Geerts, Hugo. 2009. "Of Mice and Men." *CNS Drugs* 23 (11): 915–26. doi:10.2165/11310890-000000000-00000.

Geissler, Paul Wenzel. 2013. "Public Secrets in Public Health: Knowing Not to Know While Making Scientific Knowledge." *American Ethnologist* 40 (1): 13–34. doi:10.1111/amet.12002.

Gelernter, J., S. O'Malley, N. Risch, H. R. Kranzler, J. Krystal, K. Merikangas, J. L. Kennedy, and K. K. Kidd. 1991. "No Association between an Allele at the D2 Dopamine Receptor Gene (DRD2) and Alcoholism." *JAMA* 266 (13): 1801–7. doi:10.1001/jama.1991.03470130081033.

Geller, Gail, Barbara A. Bernhardt, Mary Gardner, Joann Rodgers, and Neil A. Holtzman. 2005. "Scientists' and Science Writers' Experiences Reporting Genetic Discoveries: Toward an Ethic of Trust in Science Journalism." *Genetics in Medicine* 7 (3): 198–205. doi:10.1097/01.GIM.0000156699.78856.23.

Gerlai, Robert. 1996. "Gene-Targeting Studies of Mammalian Behavior: Is It the Mutation or the Background Genotype?" *Trends in Neurosciences* 19 (5): 177–81. doi:10.1016/S0166-2236(96)20020-7.

Gieryn, Thomas F. 2002. "Three Truth-Spots." *Journal of the History of the Behavioral Sciences* 38 (2): 113–32. doi:10.1002/jhbs.10036.

———. 2006. "City as Truth-Spot: Laboratories and Field-Sites in Urban Studies." *Social Studies of Science* 36 (1): 5–38. doi:10.1177/0306312705054526.

Glaser, Barney G., and Anselm L. Strauss. 1967. *The Discovery of Grounded Theory: Strategies for Qualitative Research*. Chicago: Aldine.

Goffman, Erving. 1955. "On Face-Work: An Analysis of Ritual Elements in Social Interaction." *Psychiatry* 18 (3): 213–31. doi:10.1521/00332747.1955.11023008.

Gottesman, Irving I., and Todd D. Gould. 2003. "The Endophenotype Concept in Psychiatry: Etymology and Strategic Intentions." *American Journal of Psychiatry* 160 (4): 636–45. doi:10.1176/appi.ajp.160.4.636.

Grant, Seth G. N., Thomas J. O'Dell, Kevin A. Karl, Paul L. Stein, Philippe Soriano, and Eric R. Kandel. 1992. "Impaired Long-Term Potentiation, Spatial Learning, and Hippocampal Development in Fyn Mutant Mice." *Science* 258 (5090): 1903–10. doi:10.1126/science.1361685.

Gregory, Jane, and Steve Miller. 1998. *Science in Public: Communication, Culture, and Credibility*. New York: Plenum.

Greiffenhagen, Christian, and Wes Sharrock. 2011. "Does Mathematics Look Certain in the Front, but Fallible in the Back?" *Social Studies of Science* 41 (6): 839–66. doi:10.1177/0306312711424789.

Gusfield, Joseph R. 1981. *The Culture of Public Problems: Drinking-Driving and the Symbolic Order*. Chicago: University of Chicago Press.

Gusterson, Hugh. 2008. "Nuclear Futures: Anticipatory Knowledge, Expert Judgment, and the Lack That Cannot Be Filled." *Science and Public Policy* 35 (8): 551–60. doi:10.3152/030234208X370639.

Hagenbuch, Niels, Joram Feldon, and Benjamin K. Yee. 2006. "Use of the Elevated Plus-Maze Test with Opaque or Transparent Walls in the Detection of Mouse Strain Differences and the Anxiolytic Effects of Diazepam." *Behavioural Pharmacology* 17 (1): 31–41. doi:10.1097/01.fbp.0000189811.77049.3e.

Hall, Wayne, Adrian Carter, and Cynthia Forlini. 2015. "The Brain Disease Model of Addiction: Is It Supported by the Evidence and Has It Delivered on Its Promises?" *Lancet Psychiatry* 2 (1): 105–10. doi:10.1016/S2215-0366(14)00126-6.

Haller, Jozsef, and Mano Alicki. 2012. "Current Animal Models of Anxiety, Anxiety Disorders, and Anxiolytic Drugs." *Current Opinion in Psychiatry* 25 (1): 59–64. doi:10.1097/YCO.0b013e32834de34f.

Haraway, Donna J. 1989. *Primate Visions: Gender, Race, and Nature in the World of Modern Science*. New York: Routledge.

———. 1997. *Modest_Witness@Second_Millennium:FemaleMan_Meets_OncoMouse*. New York: Routledge.

Hayden, Erika. 2010. "Human Genome at Ten: Life Is Complicated." *Nature News* 464 (7289): 664–67. doi:10.1038/464664a.

Hedgecoe, Adam. 2001. "Schizophrenia and the Narrative of Enlightened Geneticization." *Social Studies of Science* 31 (6): 875–911. doi:10.1177/030631201031006004.

———. 2010. "Bioethics and the Reinforcement of Socio-Technical Expectations." *Social Studies of Science* 40 (2): 163–86. doi:10.1177/0306312709349781.

Hedgecoe, Adam, and Paul Martin. 2003. "The Drugs Don't Work: Expectations and the Shaping of Pharmacogenetics." *Social Studies of Science* 33 (3): 327–64. doi:10.1177/03063127030333002.

Herrnstein, Richard J., and Charles A. Murray. 1994. *The Bell Curve: Intelligence and Class Structure in American Life*. New York: Free Press.

Heurteaux, Catherine, Guillaume Lucas, Nicolas Guy, Malika El Yacoubi, Susanne Thümmler, Xiao-Dong Peng, Florence Noble, Nicolas Blondeau, Catherine Widmann, Marc Borsotto, Gabriella Gobbi, Jean-Marie Vaugeois, Guy Debonnel, and Michel Lazdunski. 2006. "Deletion of the Background Potassium Channel TREK-1 Results in a Depression-Resistant Phenotype." *Nature Neuroscience* 9 (9): 1134–41. doi:10.1038/nn1749.

Hilgartner, Stephen. 1990. "The Dominant View of Popularization: Conceptual Problems, Political Uses." *Social Studies of Science* 20 (3): 519–39. doi:10.1177/030631290020003006.

———. 2011. "Staging High-Visibility Science: Media Orientation in Genome Research." In *The Sciences' Media Connection: Public Communication and Its Repercussions*, edited by Simone Rodder, Martina Franzen, and Peter Weingart, 189–215. Dordrecht: Springer Netherlands.

———. 2017. *Reordering Life: Knowledge and Control in the Genomics Revolution*. Cambridge, MA: MIT Press.

Hogg, Sandy. 1996. "A Review of the Validity and Variability of the Elevated Plus-Maze as an Animal Model of Anxiety." *Pharmacology, Biochemistry, and Behavior* 54 (1): 21–30. doi:10.1016/0091-3057(95)02126-4.

Holden, Constance. 2008. "Parsing the Genetics of Behavior." *Science* 322 (5903): 892–95. doi:10.1126/science.322.5903.892.

Horwitz, Allan V. 2005. "Media Portrayals and Health Inequalities: A Case Study of Characterizations of Gene Environment Interactions." *The Journals of Gerontology. Series B, Psychological Sciences and Social Sciences* 60 (Special Issue 2): 48–52. doi:10.1093/geronb/60.Special_Issue_2.S48.

Huber, Lara, and Lara K. Keuck. 2013. "Mutant Mice: Experimental Organisms as Materialised Models in Biomedicine." *Studies in History and Philosophy of Biological and Biomedical Sciences* 44 (3): 385–91. doi:10.1016/j.shpsc.2013.03.001.

Hughes, Virginia. 2012. "Jonah Lehrer, Scientists, and the Nature of Truth." *Phenomena*, November 5. http://phenomena.nationalgeographic.com/2012/11/05/jonah-lehrer-nature-of-truth/.

Hyman, Steven. 2006. "Using Genetics to Understand Human Behavior: Promises and Risks." In *Wrestling with Behavioral Genetics: Science, Ethics and Public Conversation*, edited by Erik Parens, Audrey R. Chapman, and Nancy Press, 109–30. Baltimore: Johns Hopkins University Press.

Immen, Wallace. 1999. "Canada's Relaxed Mice Puzzling: Lab Rodents More Laid Back Than U.S. Cousins; Differing Behaviour Could Nullify Experiments." *Globe and Mail* [Calgary, AB], June 4, A6.

Izídio, G. S., D. M. Lopes, L. Spricigo, and A. Ramos. 2005. "Common Variations in the Pretest Environment Influence Genotypic Comparisons in Models of Anxiety." *Genes, Brain and Behavior* 4 (7): 412–19. doi:10.1111/j.1601-183X.2005.00121.x.

Jacoby, Russell, and Naomi Glauberman. 1995. *The Bell Curve Debate*. New York: Three Rivers.

Jellinek, E. Morton. 1960. *The Disease Concept of Alcoholism*. New Haven, CT: College and University Press.

Jensen, Arthur R. 1969. "How Much Can We Boost IQ and Scholastic Achievement?" *Harvard Educational Review* 39 (1): 1–123.

Jones, Jacqui. 2011. "Genetic Link to Schizophrenia." *Newcastle* [Australia] *Herald*, September 19, 9.

Jordan, Kathleen, and Michael E. Lynch. 1992. "The Sociology of a Genetic Engineering Technique: Ritual and Rationality in the Performance of the 'Plasmid Prep.'" In *The Right Tools for the Job: At Work in Twentieth-Century Life Science*, edited by Adele E. Clarke and Joan H. Fujimura, 77–104. Princeton, NJ: Princeton University Press.

Kandel, Eric R. 2007. *In Search of Memory: The Emergence of a New Science of Mind*. New York: Norton.

Keller, Evelyn Fox. 1983. *A Feeling for the Organism: The Life and Work of Barbara McClintock*. San Francisco: Freeman.

———. 1997. "Developmental Biology as a Feminist Cause?" *Osiris* 12: 16–28. doi:10.2307/301896.

———. 2000. "Models of and Models For: Theory and Practice in Contemporary Biology." *Philosophy of Science* 67 (supplement): S72–S86. doi:10.2307/188659.

———. 2010. *The Mirage of a Space between Nature and Nurture*. Durham, NC: Duke University Press.

———. 2015. "The Postgenomic Genome." In *Postgenomics: Perspectives on Biology after the*

Genome, edited by Sarah S. Richardson and Hallam Stevens, 9–31. Durham, NC: Duke University Press.

Kendler, Kenneth S. 2006. "Reflections on the Relationship between Psychiatric Genetics and Psychiatric Nosology." *American Journal of Psychiatry* 163 (7): 1138–46. doi:10.1176/ajp.2006.163.7.1138.

Kidd, Kenneth K. 1993. "Associations of Disease with Genetic Markers: Déjà vu All over Again." *American Journal of Medical Genetics* 48 (2): 71–73. doi:10.1002/ajmg.1320480202.

Kirk, Robert G. W. 2009. "Between the Clinic and the Laboratory: Ethology and Pharmacology in the Work of Michael Robin Alexander Chance, c. 1946–1964." *Medical History* 53 (4): 513–36.

———. 2014. "The Invention of the 'Stressed Animal' and the Development of a Science of Animal Welfare, 1947–86." In *Stress, Shock, and Adaptation in the Twentieth Century*, edited by David Cantor and Edmund Ramsden, 241–63. Rochester, NY: University of Rochester Press.

Knorr, Karin D. 1981. *The Manufacture of Knowledge: An Essay on the Constructivist and Contextual Nature of Science*. Oxford: Pergamon.

Knorr-Cetina, Karin D. 1999. *Epistemic Cultures: How the Sciences Make Knowledge*. Cambridge, MA: Harvard University Press.

Kohler, Robert E. 1994. *Lords of the Fly: Drosophila Genetics and the Experimental Life*. Chicago: University of Chicago Press.

———. 2002. "Place and Practice in Field Biology." *History of Science: An Annual Review of Literature, Research and Teaching* 40 (128 Pt 2): 189–210. doi:10.1177/007327530204000204.

Kolata, Gina. 1995. "Will Alcoholic Mice Teach Scientists about Human Behavior?" *New York Times*, December 26, C12.

Kurt, Mehmet, Ali Cezmi Arik, and Suleyman Çelik. 2000. "The Effects of Sertraline and Fluoxetine on Anxiety in the Elevated Plus-Maze Test in Mice." *Journal of Basic and Clinical Physiology and Pharmacology* 11 (2): 173–80. doi:10.1515/JBCPP.2000.11.2.173.

Kvaale, Erlend P., Nick Haslam, and William H. Gottdiener. 2013. "The 'Side Effects' of Medicalization: A Meta-Analytic Review of How Biogenetic Explanations Affect Stigma." *Clinical Psychology Review* 33 (6): 782–94. doi:10.1016/j.cpr.2013.06.002.

LaFollette, Marcel C. 1990. *Making Science Our Own: Public Images of Science, 1910–1955*. Chicago: University of Chicago Press.

Landecker, Hannah. 2013. "When the Control Becomes the Experiment." *Limn* 3. http://limn.it/when-the-control-becomes-the-experiment/.

Latour, Bruno. 1987. *Science in Action: How to Follow Scientists and Engineers through Society*. Cambridge, MA: Harvard University Press.

———. 2004. "Why Has Critique Run Out of Steam? From Matters of Fact to Matters of Concern." *Critical Inquiry* 30 (2): 225–48. doi:10.1086/421123.

Latour, Bruno, and Steve Woolgar. 1979. *Laboratory Life: The Construction of Scientific Facts*. Thousand Oaks, CA: Sage.

———. 1986. *Laboratory Life: The Construction of Scientific Facts*. 2nd ed. Princeton, NJ: Princeton University Press.

Lederer, Susan E. 1992. "Political Animals: The Shaping of Biomedical Research Literature in Twentieth-Century America." *Isis* 83 (1): 61–79. doi:10.1086/356025.

Lee McFarling, Usha. 1998. "Brain Chemical Tied to Anxiety May Have Role in Alcoholism." *Philadelphia Inquirer*, November 26, A03.

Lehrer, Jonah. 2010. "The Truth Wears Off: Is There Something Wrong with the Scientific Method?" *New Yorker*, December 13, 52–57.

Lemov, Rebecca. 2005. *World as Laboratory: Experiments with Mice, Mazes, and Men*. New York: Hill & Wang.

Lerner, Richard M. 2006. "Another Nine-Inch Nail for Behavioral Genetics!" *Human Development* 49 (6): 336–42. doi:10.1159/000096532.

Levin, Nadine. 2014. "Multivariate Statistics and the Enactment of Metabolic Complexity." *Social Studies of Science* 44 (4): 555–78. doi:10.1177/0306312714524845.

Levine, Harry G. 1978. "The Discovery of Addiction: Changing Conceptions of Habitual Drunkenness in America." *Journal of Studies on Alcohol* 39 (1): 143–74. doi:10.15288/jsa.1978.39.143.

Lewin, Kurt. 1935. *A Dynamic Theory of Personality*. New York: McGraw-Hill.

Lewis, Jamie, Paul Atkinson, Jean Harrington, and Katie Featherstone. 2013. "Representation and Practical Accomplishment in the Laboratory: When Is an Animal Model Good-Enough?" *Sociology* 47 (4): 776–92. doi:10.1177/0038038512457276.

Lewontin, Richard C. 1991. *Biology as Ideology: The Doctrine of DNA*. New York: Harper Perennial.

Lewontin, Richard, Steven Rose, and Leon Kamin. 1984. *Not in Our Genes: Biology, Ideology, and Human Nature*. New York: Pantheon.

Lippman, Abby. 1992. "Led (Astray) by Genetic Maps: The Cartography of the Human Genome and Health Care." *Social Science & Medicine* 35 (12): 1469–76. doi:10.1016/0277-9536(92)90049-V.

Lister, Richard G. 1987. "The Use of a Plus-Maze to Measure Anxiety in the Mouse." *Psychopharmacology* 92 (2): 180–85. doi:10.1007/BF00177912.

Lock, Margaret M. 1993. *Encounters with Aging: Mythologies of Menopause in Japan and North America*. Berkeley: University of California Press.

Lock, Margaret, Julia Freeman, Rosemary Sharples, and Stephanie Lloyd. 2006. "When It Runs in the Family: Putting Susceptibility Genes in Perspective." *Public Understanding of Science* 15 (3): 277–300. doi:10.1177/0963662506059259.

Logan, Cheryl A. 2002. "Before There Were Standards: The Role of Test Animals in the Production of Empirical Generality in Physiology." *Journal of the History of Biology* 35 (2): 329–63. doi:10.1023/A:1016036223348.

Longino, Helen E. 2013. *Studying Human Behavior: How Scientists Investigate Aggression and Sexuality*. Chicago: University of Chicago Press.

Lynch, Michael E. 1985. *Art and Artifact in Laboratory Science: A Study of Shop Work and Shop Talk in a Research Laboratory*. London: Routledge Kegan & Paul.

———. 1988. "Sacrifice and the Transformation of the Animal Body into a Scientific Object:

Laboratory Culture and Ritual Practice in the Neurosciences." *Social Studies of Science* 18 (2): 265–89. doi:10.1177/030631288018002004.

McClearn, G. E., and David A. Rodgers. 1959. "Differences in Alcohol Preference among Inbred Strains of Mice." *Quarterly Journal of Studies on Alcohol* 20: 691–95.

McGill University. 2006. "Ever-Happy Mice May Hold Key to New Treatment of Depression." *Science Daily: Science News*, August 23. http://www.sciencedaily.com/releases/2006/08/060822180641.htm.

McGuffin, Peter, Brien Riley, and Robert Plomin. 2001. "Toward Behavioral Genomics." *Science* 291 (5507): 1232–49. doi:10.1126/science.1057264.

Merton, Robert K. 1973 [1942]. "The Normative Structure of Science." In *The Sociology of Science: Theoretical and Empirical Investigations*, 267–78. Chicago: University of Chicago Press.

"Mice Study Shows Genes Are Not Always Destiny." 1999. *Reuters News*, June 3.

Miller, Michael E. 2015. "Scientists May Have Found a Way to Make You Forget That You're Addicted to Meth." *Washington Post: Morning Mix*, August 7. https://www.washingtonpost.com/news/morning-mix/wp/2015/08/07/scientists-may-have-found-a-way-to-make-you-forget-that-youre-addicted-to-meth/.

Milner, Lauren C., and John C. Crabbe. 2008. "Three Murine Anxiety Models: Results from Multiple Inbred Strain Comparisons." *Genes, Brain and Behavior* 7 (4): 496–505. doi:10.1111/j.1601-183X.2007.00385.x.

Mitchell, Sandra D. 2009. *Unsimple Truths: Science, Complexity, and Policy*. Chicago: University of Chicago Press.

Mitman, Gregg, and Anne Fausto-Sterling. 1992. "Whatever Happened to Planaria? C. M. Child and the Physiology of Inheritance." In *The Right Tools for the Job: At Work in 20th Century Life Sciences*, edited by Adele E. Clarke and Joan H. Fujimura, 172–97. Princeton, NJ: Princeton University Press.

Mol, Annemarie. 2002. *The Body Multiple: Ontology in Medical Practice*. Durham, NC: Duke University Press.

Mombaerts, Peter, Alan R. Clarke, Martin L. Hooper, and Susumu Tonegawa. 1991. "Creation of a Large Genomic Deletion at the T-Cell Antigen Receptor Beta-Subunit Locus in Mouse Embryonic Stem Cells by Gene Targeting." *Proceedings of the National Academy of Sciences* 88 (8): 3084–87. doi:10.1073/pnas.88.8.3084.

Moser, Paul C. 1989. "An Evaluation of the Elevated Plus-Maze Test Using the Novel Anxiolytic Buspirone." *Psychopharmacology* 99 (1): 48–53. doi:10.1007/BF00634451.

Murray, Fiona. 2010. "The Oncomouse That Roared: Hybrid Exchange Strategies as a Source of Distinction at the Boundary of Overlapping Institutions." *American Journal of Sociology* 116 (2): 341–88. doi:10.1086/653599.

Myers, Natasha. 2015. *Rendering Life Molecular: Models, Modelers, and Excitable Matter*. Durham, NC: Duke University Press.

Naimi, Timothy S., Robert D. Brewer, Ali Mokdad, Clark Denny, Mary K. Serdula, and James S. Marks. 2003. "Binge Drinking Among US Adults." *JAMA* 289 (1): 70–75. doi:10.1001/jama.289.1.70.

National Institute on Alcohol Abuse and Alcoholism, ed. 2004. "NIAAA Council Approves

Definition of Binge Drinking." *NIAAA Newsletter* 3 (Winter): 3. http://pubs.niaaa.nih.gov/publications/Newsletter/winter2004/Newsletter_Number3.pdf.

National Institute on Drug Abuse (NIDA). 2007. "Drugs, Brains, and Behavior: The Science of Addiction." Bethesda, MD: National Institutes of Health.

Nelkin, Dorothy. 1987. *Selling Science: How the Press Covers Science and Technology*. New York: Freeman.

Nelkin, Dorothy, and M. Susan Lindee. 1995. *The DNA Mystique: The Gene as a Cultural Icon*. New York: Freeman.

Nelson, Nicole. 2013. "Modeling Mouse, Human, and Discipline: Epistemic Scaffolds in Animal Behavior Genetics." *Social Studies of Science* 43 (1): 3–29. doi:10.1177/0306312712463815.

———. 2015. "A Knockout Experiment: Disciplinary Divides and Experimental Skill in Animal Behavior Genetics." *Medical History* 59 (3): 465–85. doi:10.1017/mdh.2015.30.

Niaz, Mansoor. 2015. "Myth 19: That the Millikan Oil-Drop Experiment Was Simple and Straightforward." In *Newton's Apple and Other Myths about Science*, edited by Ronald Numbers and Kostas Kampourakis, 157–63. Cambridge, MA: Harvard University Press.

Niewöhner, Jörg. 2011. "Epigenetics: Embedded Bodies and the Molecularisation of Biography and Milieu." *BioSocieties* 6 (3): 279–98. doi:10.1057/biosoc.2011.4.

Nuffield Council on Bioethics. 2002. *Genetics and Human Behaviour: The Ethical Context*. London: Nuffield Council on Bioethics.

Ogle, Andy. 2002. "Unlocking Mice Traits: Results Could Lead to Answers for Our Behaviour." *Edmonton Journal*, August 6, B3.

Paigen, Kenneth, and Janan T. Eppig. 2000. "A Mouse Phenome Project." *Mammalian Genome* 11 (9): 715–17. doi:10.1007/s003350010152.

Panlab. 2010. "Integrated Solutions Anxiety." http://www.panlab.com/panlabWeb/Solution/php/displaySol.php?nameSolution=ANXIETY#contenido1.

Panofsky, Aaron L. 2007. "Fielding Controversy: The Genesis and Structure of Behavior Genetics." PhD diss., New York University.

———. 2011. "Field Analysis and Interdisciplinary Science: Scientific Capital Exchange in Behavior Genetics." *Minerva* 49 (3): 295–316. doi:10.1007/s11024-011-9175-1.

———. 2014. *Misbehaving Science: Controversy and the Development of Behavior Genetics*. Chicago: University of Chicago Press.

———. 2015. "From Behavior Genetics to Postgenomics." In *Postgenomics: Perspectives on Biology after the Genome*, edited by Sarah S. Richardson and Hallam Stevens, 150–73. Durham, NC: Duke University Press.

Paul, Diane B. 1998. "The Rockefeller Foundation and the Origins of Behavior Genetics." In *The Politics of Heredity: Essays on Eugenics, Biomedicine, and the Nature-Nurture Debate*, 53–80. Albany: State University of New York Press.

Pellow, Sharon, Philippe Chopin, Sandra E. File, and Mike Briley. 1985. "Validation of Open: Closed Arm Entries in an Elevated Plus-Maze as a Measure of Anxiety in the Rat." *Journal of Neuroscience Methods* 14 (3): 149–67. doi:10.1016/0165-0270(85)90031-7.

Pfaff, Donald. 2001. "Precision in Mouse Behavior Genetics." *Proceedings of the National Academy of Sciences* 98 (11): 5957–60. doi:10.1073/pnas.101128598.

Phelan, Jo C. 2005. "Geneticization of Deviant Behavior and Consequences for Stigma: The Case of Mental Illness." *Journal of Health and Social Behavior* 46 (4): 307–22. doi:10.1177/002214650504600401.

Picciotto, Marina R., and David W. Self. 1999. "Testing the Genetics of Behavior in Mice." *Science* 285 (5436): 2067. doi:10.1126/science.285.5436.2067d.

Pinch, Trevor J. 1985. "Towards an Analysis of Scientific Observation: The Externality and Evidential Significance of Observational Reports in Physics." *Social Studies of Science* 15 (1): 3–36. doi:10.1177/030631285015001001.

Pohorecky, Larissa A. 1999. "Testing the Genetics of Behavior in Mice." *Science* 285 (5436): 2067.

Pollack, Andrew. 2003. "Can Drugs Make Us Happier? Smarter?" *New York Times*, November 11, F12.

Proctor, Robert N. 1995. *Cancer Wars: How Politics Shapes What We Know and Don't Know about Cancer*. New York: Basic Books.

Proctor, Robert N., and Londa L. Schiebinger, eds. 2008. *Agnotology: The Making and Unmaking of Ignorance*. Stanford, CA: Stanford University Press.

Rader, Karen. 2004. *Making Mice: Standardizing Animals for American Biomedical Research, 1900–1955*. Princeton, NJ: Princeton University Press.

Ramboz, Sylvie, Frédéric Saudou, Djamel Aït Amara, Catherine Belzung, Louis Segu, René Misslin, Marie-Christine Buhot, and René Hen. 1995. "5-HT1B Receptor Knock Out: Behavioral Consequences." *Behavioural Brain Research* 73 (12): 305–12. doi:10.1016/0166-4328(96)00119-2.

Ramos, André. 2008. "Animal Models of Anxiety: Do I Need Multiple Tests?" *Trends in Pharmacological Sciences* 29 (10): 493–98. doi:10.1016/j.tips.2008.07.005.

Ramsden, Edmund. 2011. "From Rodent Utopia to Urban Hell: Population, Pathology, and the Crowded Rats of NIMH." *Isis* 102 (4): 659–88. doi:10.1086/663598.

———. 2015. "Making Animals Alcoholic: Shifting Laboratory Models of Addiction." *Journal of the History of the Behavioral Sciences* 51 (2): 164–94. doi:10.1002/jhbs.21715.

Reynolds, Gretchen. 2013. "How Exercise Can Calm Anxiety." *New York Times: Well Blog*. http://well.blogs.nytimes.com/2013/07/03/how-exercise-can-calm-anxiety/.

Rheinberger, Hans-Jörg. 1997. *Toward a History of Epistemic Things: Synthesizing Proteins in the Test Tube*. Stanford, CA: Stanford University Press.

Richards, Martin. 2006. "Heredity: Lay Understanding." In *Living with the Genome: Ethical and Social Aspects of Human Genetics*, edited by Angus Clarke and Flo Ticehurst, 177–82. New York: Palgrave Macmillan.

Richardson, Sarah S. 2015. "Maternal Bodies in the Postgenomic Order: Gender and the Explanatory Landscape of Epigenetics." In *Postgenomics: Perspectives on Biology after the Genome*, edited by Sarah S. Richardson and Hallam Stevens, 210–31. Durham, NC: Duke University Press.

Room, Robin. 1974. "Governing Images and the Prevention of Alcohol Problems." *Preventive Medicine* 3 (1): 11–23. doi:10.1016/0091-7435(74)90059-0.

———. 1983. "Sociological Aspects of the Disease Concept of Alcoholism." *Research Advances in Alcohol and Drug Problems* 7: 47–91.

Rose, Nikolas. 2007. *The Politics of Life Itself: Biomedicine, Power, and Subjectivity in the Twenty-First Century*. Princeton, NJ: Princeton University Press.

Rose, Nikolas, and Joelle M. Abi-Rached. 2013. *Neuro: The New Brain Sciences and the Management of the Mind*. Princeton, NJ: Princeton University Press.

Rose, Steven. 1997. *Lifelines: Biology beyond Determinism*. Oxford: Oxford University Press.

Rosenthal, Robert. 1966. *Experimenter Effects in Behavioral Research*. New York: Appleton-Century-Crofts.

Rosenthal, Robert, and Kermit L. Fode. 1963. "The Effect of Experimenter Bias on the Performance of the Albino Rat." *Behavioral Science* 8 (3): 183–89. doi:10.1002/bs.3830080302.

Rosoff, Philip M. 2010. "In Search of the Mommy Gene: Truth and Consequences in Behavioral Genetics." *Science, Technology & Human Values* 35 (2): 200–43. doi:10.1177/0162243909340260.

Sacks, Harvey, Emanuel A. Schegloff, and Gail Jefferson. 1974. "A Simplest Systematics for the Organization of Turn-Taking for Conversation." *Language* 50 (4): 696–735. doi:10.2307/412243.

Schegloff, Emanuel A., Gail Jefferson, and Harvey Sacks. 1977. "The Preference for Self-Correction in the Organization of Repair in Conversation." *Language* 53 (2): 361–82.

Schneider, Joseph W. 1978. "Deviant Drinking as Disease: Alcoholism as a Social Accomplishment." *Social Problems* 25 (4): 361–72. doi:10.2307/800489.

Schwartz, John. 1995. "Genetics: Why Some Mice May Be More Timid." *Washington Post*, September 11, A02.

Schwartzberg, Pamela L., Alan M. Stall, Jeff D. Hardin, Katherine S. Bowdish, Teresa Humaran, Sharon Boast, Margaret L. Harbison, Elizabeth J. Robertson, and Stephen P. Goff. 1991. "Mice Homozygous for the Ablm1 Mutation Show Poor Viability and Depletion of Selected B and T Cell Populations." *Cell* 65 (7): 1165–75. doi:10.1016/0092-8674(91)90012-N.

"Science Notebook: Mice with High Anxiety." 1999. *Washington Post*, October 11, A11.

Shapin, Steven, and Simon Schaffer. 1989. *Leviathan and the Air-Pump: Hobbes, Boyle, and the Experimental Life*. Princeton, NJ: Princeton University Press.

Shepherd, Jon K., Savraj S. Grewal, Allan Fletcher, David J. Bill, and Colin T. Dourish. 1994. "Behavioural and Pharmacological Characterisation of the Elevated 'Zero-Maze' as an Animal Model of Anxiety." *Psychopharmacology* 116 (1): 56–64. doi:10.1007/BF02244871.

Shostak, Sara. 2007. "Translating at Work: Genetically Modified Mouse Models and Molecularization in the Environmental Health Sciences." *Science, Technology & Human Values* 32 (3): 315–38. doi:10.1177/0162243906298353.

———. 2013. *Exposed Science: Genes, the Environment, and the Politics of Population Health*. Berkeley: University of California Press.

Silva, Alcino J., Richard Paylor, Jeanne M. Wehner, and Susumu Tonegawa. 1992. "Impaired Spatial Learning in α-Calcium-Calmodulin Kinase II Mutant Mice." *Science* 257 (5067): 206–11. doi:10.1126/science.1321493.

Silva, M. T., C. R. Alves, and E. M. Santarem. 1999. "Anxiogenic-Like Effect of Acute and Chronic Fluoxetine on Rats Tested on the Elevated Plus-Maze." *Brazilian Journal of Medical and Biological Research* 32 (3): 333–39. doi:10.1590/S0100-879X1999000300014.

Sismondo, Sergio. 1997. "Deflationary Metaphysics and the Construction of Laboratory Mice." *Metaphilosophy* 28 (3): 219–32. doi:10.1111/1467-9973.00051.

———. 2008. "Science and Technology Studies and an Engaged Program." In *The Handbook of Science and Technology Studies*, 3rd ed., edited by Edward J. Hackett, Olga Amsterdamska, Michael Lynch, and Judy Wajcman, 13–31. Cambridge, MA: MIT Press.

———. 2009. *An Introduction to Science and Technology Studies*. 2nd ed. Chichester, UK: Wiley-Blackwell.

Soriano, Philippe, Charles Montgomery, Robert Geske, and Allan Bradley. 1991. "Targeted Disruption of the c-Src Proto-Oncogene Leads to Osteopetrosis in Mice." *Cell* 64 (4): 693–702. doi:10.1016/0092-8674(91)90499-O.

Star, Susan Leigh. 1985. "Scientific Work and Uncertainty." *Social Studies of Science* 15 (3): 391–427. doi:10.1177/030631285015003001.

Stevens, Hallam. 2011. "On the Means of Bio-Production: Bioinformatics and How to Make Knowledge in a High-Throughput Genomics Laboratory." *BioSocieties* 6 (2): 217–42. doi:10.1057/biosoc.2010.38.

Stevens, Hallam, and Sarah S. Richardson. 2015. "Beyond the Genome." In *Postgenomics: Perspectives on Biology after the Genome*, edited by Sarah S. Richardson and Hallam Stevens, 1–8. Durham, NC: Duke University Press.

Stocking, Holly. 1999. "How Journalists Deal with Scientific Uncertainty." In *Communicating Uncertainty: Media Coverage of New and Controversial Science*, edited by Sharon Friedman, Sharon Dunwoody, and Carol Rogers, 23–42. New York: Routledge.

Svendsen, Mette N., and Lene Koch. 2013. "Potentializing the Research Piglet in Experimental Neonatal Research." *Current Anthropology* 54 (S7): S118–S128. doi:10.1086/671060.

Tabery, James. 2014. *Beyond Versus: The Struggle to Understand the Interaction of Nature and Nurture*. Cambridge, MA: MIT Press.

"The Obesity Gene." 1994. *New York Times*, December 5, A18.

Thiele, Todd E., Donald J. Marsh, Linda Ste. Marie, Ilene L. Bernstein, and Richard D. Palmiter. 1998. "Ethanol Consumption and Resistance Are Inversely Related to Neuropeptide Y Levels." *Nature* 396 (6709): 366–69. doi:10.1038/24614.

Thomas, Kirk R., and Mario R. Capecchi. 1990. "Targeted Disruption of the Murine Int-1 Proto-Oncogene Resulting in Severe Abnormalities in Midbrain and Cerebellar Development." *Nature* 346 (6287): 847–50. doi:10.1038/346847a0.

Tordoff, Michael G., Alexander A. Bachmanov, Mark I. Friedman, and Gary K. Beauchamp. 1999. "Testing the Genetics of Behavior." *Science* 285 (5436): 2068–69.

Traweek, Sharon. 1992. *Beamtimes and Lifetimes: The World of High Energy Physicists*. Cambridge, MA: Harvard University Press.

Tutton, Richard. 2011. "Promising Pessimism: Reading the Futures to Be Avoided in Biotech." *Social Studies of Science* 41 (3): 411–29. doi:10.1177/0306312710397398.

Underage Drinking Enforcement Training Center. 2005. "Drinking in America: Myths, Realities, and Prevention Policy." Office of Juvenile Justice and Delinquency Prevention. http://web.archive.org/web/20150910124238/http://www.udetc.org/documents/Drinking_in_America.pdf.

Van Der Staay, F. J., and T. Steckler. 2002. "The Fallacy of Behavioral Phenotyping without Standardisation." *Genes, Brain and Behavior* 1 (1): 9–13. doi:10.1046/j.1601-1848.2001.00007.x.

Venter, J. Craig. 2007. *A Life Decoded: My Genome, My Life*. New York: Viking.

Vertesi, Janet. 2015. *Seeing Like a Rover: How Robots, Teams, and Images Craft Knowledge of Mars*. Chicago: University of Chicago Press.

Volkow, Nora D., and George Koob. 2015. "Brain Disease Model of Addiction: Why Is It So Controversial?" *Lancet Psychiatry* 2 (8): 677–79. doi:10.1016/S2215-0366(15)00236-9.

Vrecko, Scott. 2010a. "Birth of a Brain Disease: Science, the State and Addiction Neuropolitics." *History of the Human Sciences* 23 (4): 52–67. doi:10.1177/0952695110371598.

———. 2010b. "'Civilizing Technologies' and the Control of Deviance." *BioSocieties* 5 (1): 36–51. doi:10.1057/biosoc.2009.8.

Wade, Nicholas. 2008. "Schizophrenia Studies Find New Hurdles." *New York Times*, July 31, A18.

———. 2010. "A Decade Later, Genetic Map Yields Few New Cures." *New York Times*, June 12, sec. Health/Research. http://www.nytimes.com/2010/06/13/health/research/13genome.html.

Wahlsten, Douglas. 1994a. "The Intelligence of Heritability." *Canadian Psychology/Psychologie Canadienne* 35 (3): 244–60. doi:10.1037/0708-5591.35.3.244.

———. 1994b. "Nascent Doubts May Presage Conceptual Clarity: Reply to Surbey." *Canadian Psychology/Psychologie Canadienne* 35 (3): 265–67. doi:10.1037/0708-5591.35.3.265.

———. 2003. "Airbrushing Heritability." *Genes, Brain, and Behavior* 2 (6): 327–29, discussion 330–31.

Wahlsten, Douglas, Pamela Metten, Tamara J. Phillips, Stephen L. Boehm II, Sue Burkhart-Kasch, Janet Dorow, Sharon Doerksen et al. 2003. "Different Data from Different Labs: Lessons from Studies of Gene-Environment Interaction." *Journal of Neurobiology* 54 (1): 283–311. doi:10.1002/neu.10173.

Walf, Alicia A., and Cheryl A. Frye. 2007. "The Use of the Elevated Plus Maze as an Assay of Anxiety-Related Behavior in Rodents." *Nature Protocols* 2 (2): 322–28. doi:10.1038/nprot.2007.44.

———. 2009. "Using the Elevated Plus Maze as a Bioassay to Assess the Effects of Naturally Occurring and Exogenously Administered Compounds to Influence Anxiety-Related Behaviors of Mice." In *Mood and Anxiety Related Phenotypes in Mice: Characterization Using Behavioral Tests*, edited by Todd D. Gould, 225–46. New York: Humana.

Wall, Philip M., and Claude Messier. 2000. "Ethological Confirmatory Factor Analysis of Anxiety-Like Behaviour in the Murine Elevated Plus-Maze." *Behavioural Brain Research* 114 (1–2): 199–212. doi:10.1016/S0166-4328(00)00229-1.

Walton, Dawn. 2005. "Study Turns Pot Wisdom on Its Head." *Globe and Mail* [Calgary, AB], October 14, A1.

Williams, Richard, Jackie E. Lim, Bettina Harr, Claudia Wing, Ryan Walters, Margaret G. Distler, Meike Teschke et al. 2009. "A Common and Unstable Copy Number Variant Is Associated with Differences in Glo1 Expression and Anxiety-Like Behavior." *PLoS ONE* 4 (3): e4649. doi:10.1371/journal.pone.0004649.

Wimsatt, William C., and James R. Griesemer. 2007. "Reproducing Entrenchments to Scaffold Culture: The Central Role of Development in Cultural Evolution." In *Integrating Evolution and Development: From Theory to Practice*, edited by Roger Sansom and Robert N. Brandon, 227–323. Cambridge, MA: MIT Press.

Wolfer, David P., Oxana Litvin, Samuel Morf, Roger M. Nitsch, Hans-Peter Lipp, and Hanno Würbel. 2004. "Laboratory Animal Welfare: Cage Enrichment and Mouse Behaviour." *Nature* 432 (7019): 821–22. doi:10.1038/432821a.

Würbel, Hanno. 2000. "Behaviour and the Standardization Fallacy." *Nature Genetics* 26 (3): 263–263. doi:10.1038/81541.

———. 2001. "Ideal Homes? Housing Effects on Rodent Brain and Behaviour." *Trends in Neurosciences* 24 (4): 207–11. doi:10.1016/S0166-2236(00)01718-5.

———. 2002. "Behavioral Phenotyping Enhanced: Beyond (Environmental) Standardization." *Genes, Brain and Behavior* 1 (1): 3–8. doi:10.1046/j.1601-1848.2001.00006.x.

Würbel, Hanno, Rosemary Chapman, and Craig Rutland. 1998. "Effect of Feed and Environmental Enrichment on Development of Stereotypic Wire-Gnawing in Laboratory Mice." *Applied Animal Behaviour Science* 60 (1): 69–81. doi:10.1016/S0168-1591(98)00150-6.

Würbel, Hanno, Markus Stauffacher, and Dietrich von Holst. 1996. "Stereotypies in Laboratory Mice—Quantitative and Qualitative Description of the Ontogeny of 'Wire-Gnawing' and 'Jumping' in Zur:ICR and Zur:ICR Nu." *Ethology* 102 (3): 371–85. doi:10.1111/j.1439-0310.1996.tb01133.x.

Wynne, Brian. 2005. "Reflexing Complexity: Post-Genomic Knowledge and Reductionist Returns in Public Science." *Theory, Culture & Society* 22 (5): 67–94. doi:10.1177/0263276405057192.

Xu, Ming, Rosario Moratalla, Lisa H. Gold, Noboru Hiroi, George F. Koob, Ann M. Graybiel, and Susumu Tonegawa. 1994. "Dopamine D1 Receptor Mutant Mice Are Deficient in Striatal Expression of Dynorphin and in Dopamine-Mediated Behavioral Responses." *Cell* 79 (4): 729–42. doi:10.1016/0092-8674(94)90557-6.

Yeoman, Barry. 2003. "Can We Trust Research Done with Lab Mice?" *Discover Magazine*, July 1, 64–71.

Zehr, Stephen C. 1999. "Scientists' Representations of Uncertainty." In *Communicating Uncertainty: Media Coverage of New and Controversial Science*, edited by Sharon Friedman, Sharon Dunwoody, and Carol Rogers, 3–22. New York: Routledge.

Zhuang, Xiaoxi, Cornelius Gross, Luca Santarelli, Valerie Compan, Anne-Cécile Trillat, and René Hen. 1999. "Altered Emotional States in Knockout Mice Lacking 5-HT1A or 5-HT1B Receptors." *Neuropsychopharmacology* 21 (2 Suppl): 52S–60S. doi:10.1016/S0893-133X(99)00047-0.

INDEX

Page numbers followed by "f" indicate figures.